EXTRAORDINARY POPULAR DELUSIONS
OF OUR TIMES

A Modern Sequel to Charles Mackay's 1841 Literary Classic
"Extraordinary Popular Delusions and the Madness of Crowds"

DANIEL MARTIN

EXTRAORDINARY POPULAR DELUSIONS OF OUR TIMES
FIRST EDITION

Copyright © 2024 Daniel Martin

All rights reserved. No portion of this book may be reproduced in any form or by any electronic means without written permission from the publisher or author except as permitted by U.S. copyright law.

ISBN: 979-8-9909604-0-4 (Paperback)
ISBN: 979-8-9909604-1-1 (eBook)
ISBN: 979-8-9909604-2-8 (Audio Book)
ISBN: 979-8-9909604-3-5 (Hardcover)
ISBN: 979-8-9909604-4-2 (Large Print)

Published by:
Wine Knot, Inc.
1073 Falls Curve
Chaska, Minnesota, 55318

Printed in the United States of America

www.popular-delusions.com

*Dedicated to Georgiann, Ted, Ben, and Sammy,
who regularly disabuse me of my delusions.*

As with any true classic, once it is read, it is hard to imagine not having known of it—and there is a compulsion to recommend it to others.
—**ANTHONY TOBIAS**, in his foreword to the 1980 Harmony Brooks edition
of "Extraordinary Popular Delusions and the Madness of Crowds"

CONTENTS

Preface ix

VOLUME 1: YOUR LIFE (HEALTH AND WELLNESS DELUSIONS)
1. The Anti-Vaccine Movement 3
2. Alternative Medicine 8
3. Chiropractic 11
4. Homeopathy 15
5. Naturopathy 18
6. Acupuncture 21
7. Essential Oils 24
8. Dietary Supplements 27
9. Detox Schemes 31
10. Cupping 36
11. Ear Candling 38
12. Reiki 40
13. Alternative Medicine for Pets 42
14. Fad Diets 45
15. Biodynamics 48
16. Monsanto and the GMO Delusion 51

VOLUME 2: YOUR MONEY (FINANCIAL DELUSIONS)
17 Dot-Com Bubble 57
18 U.S. Housing Bubble and Global Financial Crisis (2007-08) 63
19 Ghost Cities—The Chinese Housing Bubble 67
20 Cryptocurrency—The Reexamination of Money 70
21 The Bernie Madoff Ponzi Scheme 79
22 Whole Lotto Delusion 83
23 Multi-Level Marketing 86
24 Timeshares 90

VOLUME 3: YOUR SOUL (CULTURAL, RELIGIOUS, AND POLITICAL DELUSIONS)
Cultural Delusions
25 Conspiracy Theories—An Overview 96
26 The Mother of all Conspiracy Theories 99
27 The Sovereign Citizen Movement 101
28 Holocaust Denial 105
29 Unidentified Flying Objects 107
30 Flat Earth 110
31 HIV/AIDS Denialism 113
32 Climate Change Denialism 118
33 Coronavirus Conspiracy Theories 121
34 Year 2000 Bug (Y2K) 124
35 Nessie, Bigfoot, and the Abominable Snowman 127
36 Chemtrails 131
37 The Birther Movement 133
38 Psychic Services 136
39 Alex Jones and the Sandy Hook School Shooting 143
40 Catfishing 146
41 Conspiracy Theory Satire—Birds Aren't Real 148

Religious Delusions
42 Mainstream Religion 151
43 Televangelism 153

44	Faith Healing	158
45	Serpent Handling	162
46	Catholic Priest Sex Abuse Cover-up	165
47	Church of Jesus Christ of Latter-Day Saints	169
48	Jehovah's Witnesses	173
49	Church of Eckankar	175
50	Dianetics and the Church of Scientology	177
51	The Unification Church (Moonies)	183
52	End Times—The Mayan Apocalypse	185
53	End Times—Heaven's Gate and the Hale-Bopp Comet	187
54	Jim Jones and the Jonestown Massacre	189
55	Kenya Starvation Cult	192
56	Lighthouse Cult	194
57	Branch Davidians	197
58	Falun Gong	200
59	Intelligent Design and the Flying Spaghetti Monster	203

Political Delusions

60	QAnon, Proud Boys, January 6 and the Trump Delusion	208
61	Dominion Voting Machines	215
62	Suicide Bombing	217
63	The Kim Dynasty of North Korea	220
64	Albania—The Hermit Kingdom	223
65	The Communist Delusion	226

Epilogue	230
Notes and Sources	233

PREFACE

Popular delusions began so early, spread so widely, and have lasted so long, that instead of two or three volumes, fifty would scarcely suffice to detail their history.
—CHARLES MACKAY

CHARLES MACKAY'S LITERARY CLASSIC "Extraordinary Popular Delusions and the Madness of Crowds" was written in 1841 and has been in print continuously since then. I first read it in 1979 while on a walkabout hitchhiking around the world. I was captivated by the "Madness of Crowds" and vowed to one day write a modern sequel. This idea has been knocking about in the deep recesses of my brain since then. I am glad to finally have the chance to write it.

You can enjoy this book without having to read the original. Human folly has always been entertaining. Bitcoin and alternative medicine can be as fascinating and as nutty as crusades and witch trials.

Despite advancements in education, scientific knowledge, and the ease of online access to the collective body of world knowledge, there may be more delusions now than 200 years ago. Fortunately, modern delusions are easy to spot because so many social media users wear them on their sleeves.

If you suspect the Illuminati intends to plant a microchip in your brain, confirming your suspicions is alarmingly easy. Just a few hours of social media research will demonstrate that the conspiracy was even more vast than you suspected, and you can quickly share your findings with like-minded acquaintances. Then, by carefully purging dissent, you can craft an echo chamber of enthusiastic reinforcement for your beliefs.

As with Mackay's original work, this book is merely a sampling from the vast miscellany of human folly, emphasizing those of recent memory. Some of Mackay's subjects (such as witch-burning and alchemy) have run their cycle of madness and are now finished. Others remain despite centuries of evidence against them. Homeopathists and astrologists ply their trades today much as they did hundreds of years ago.

I did not attempt to catalog every human folly. I merely selected a few that caught my interest. Some may not even be delusions in the truest sense. Ponzi schemes, for example, are more crime than delusion, but they do contain elements of delusion for both the perpetrator and the victim. This work will focus on popular delusions that have gained traction despite the glare of common sense and widely available and reliable contradictory information.

Like religions and conspiracy theories, popular delusions mutate, grow, disappear, and reappear continuously. Some subjects will be new to the reader, but most are likely familiar. I have endeavored to present them in a narrative form that gains some resonance from a fresh perspective on the madness we all share as part of the human condition.

PREFACE

A FEW WORDS ABOUT SOURCING

Much of the material in this book is opinion, stemming from my life history and personal experience. I worked in information technology during Y2K and vividly remember watching my servers and systems quietly do nothing as the clock ticked gently into the new millennium. I invested in telecom and internet stocks that cratered during the Dot-Com bubble. I have family and friends who harbor conspiracy theories and are fully committed to alternative medicine. My political and religious observations are informed by extensive travel to more than 100 countries, including anti-religious countries like Russia, China, Cuba, and Vietnam.

I have been a voracious reader since childhood and a news junkie my entire adult life. This work covers a wide range of topics, and while I have opinions and ideas on each, I am not a subject matter expert in every area. I supplement my experience and views by relying on source material such as books, newspaper and magazine articles, documentary movies, company websites, news and opinion websites, and yes—Gasp!—even Wikipedia, which has overcome its initial reputation for wobbly reliability to an unsurpassed resource for links to source material for nearly any subject.

My objective is not to be a compendium of facts. I prefer to tell stories about people and how they think. I have referenced (at the chapter level) many of the sources I used to provide credit where it is due and allow the curious to dig deeper into topics of interest.

I don't expect readers to agree with everything I have written. On the contrary, I skewer some sacred cows and would not be surprised if I offend some true believers, alternative medicine providers, and others. I hope that even these readers will be entertained and consider that some of their cherished beliefs may deserve a second look.

VOLUME 1

YOUR LIFE (HEALTH AND WELLNESS DELUSIONS)

Do not be so open-minded that your brains fall out.
—G.K. CHESTERTON

1

THE ANTI-VACCINE MOVEMENT

There is no vaccine against stupidity.
—ALBERT EINSTEIN

PERHAPS THE MOST HARMFUL AND DANGEROUS health delusion of our time is the anti-vaccine (anti-vax) movement. The World Health Organization places the anti-vax movement in the top ten of all global health threats.

Anti-vax refers to opposition to vaccination against contagious diseases despite access to and awareness of the vaccine. Anti-vaxxers may also refer to themselves as Vaccine Hesitant, Vaccine Skeptics, or advocates for "Medical Freedom," appropriating the abortion movement slogan "My body, My choice."

Anti-vaccine sentiment is not new. People have been afraid of vaccines since they were first invented. Anti-vaccination leagues developed in the early 1800s in England when widespread smallpox vaccination began. Various anti-vaccination societies in the United States date back to the late 1800's.

Disinformation and conspiracy theories spread by anti-vaxxers lead to fear and public debates around the need, safety, and efficacy of vaccines and vaccine mandates. However, there is no serious debate in mainstream medical and scientific communities about the benefits of vaccination while acknowledging rare but serious side effects. Millions of lives have been saved worldwide by preventing diseases such as smallpox, diphtheria, hepatitis, polio, measles, and the flu.

Vaccines have been overwhelmingly effective in reducing hospitalization and death from COVID-19. Hundreds of millions of COVID-19 vaccine doses have been administered worldwide with an astonishingly small number of side effects. Vaccines are one of the most significant accomplishments of modern medical science. The objective evidence that vaccines are safe and effective is rock solid.

Yet, people still oppose them. Deep-rooted emotional, philosophical, political, or spiritual beliefs often underlie vaccine opposition. When probed, anti-vaxxers usually base their objection on fear (of public authorities, the pharmaceutical industry, or even of needles), misinformation, lack of understanding of how vaccines work, or some conspiracy theory.

These concerns are often conflated. When one claim is debunked, they shift to another, and another, and then back to where they started, subverting rational debate. Thousands of anti-vaccine websites reinforce their confirmation bias. Many of them wander off into bizarre and non-sensical conspiracy theories. My personal favorite is the notion that microchips designed to control your behavior are hidden in vaccines. The "chips in vaccines" mindset requires a complete misunderstanding of both biology and computer science, plus a healthy dose of mistrust of the government, industry in general, and Bill Gates in particular.

1998 marked an unfortunate advance in the anti-vax movement.

British researcher Andrew Wakefield published a study in The Lancet, a prestigious medical journal, showing a link between the MMR (Measles, Mumps and Rubella) vaccine and autism. This report prompted a wave of fear about vaccines that continues to this day. The study was quickly discredited, and The Lancet retracted it in 2010, but the damage was done.

Wakefield declined to conduct a more extensive study to substantiate or refute his conclusions. Investigative journalists revealed both scientific fraud and conflicts of interest in the findings. An attorney representing parents of autistic children partially funded Wakefield's work. These parents were searching for evidence to support the claim that vaccines had damaged their children and paid Wakefield to find it.

Subsequent studies have shown no link between vaccines and autism. And yet the stark terror of an awful condition brought on by a government-mandated vaccine remains buried deep in the psyche of anti-vax parents. They refuse to let go.

Alternative medicine providers are well-represented in the anti-vaccine movement, especially chiropractors, homeopathists, and naturopaths. This is at least partly due to the early philosophies that shaped the foundation of these groups. The founder of the practice of chiropractic (David Daniel Palmer) was dogmatically opposed to what he called "filthy animal poisons" and was unwilling to consider any new scientific advancements that superseded his theory of medicine. The state of Oregon tracks vaccination rates among all licensed healthcare providers, and their data shows that chiropractors are the least vaccinated medical professionals, significantly less than the general public and far less than mainstream medical professionals.

Chiropractic trade groups like the American Chiropractic Association and the International Chiropractors Association take a neutral position on vaccinations and then slyly shift the debate on the safety and efficacy of vaccines to an issue of civil liberty. They advocate for "Medical Freedom" and claim that chiropractic principles favor the enhancement of natural immunity over artificial immunization. They oppose compulsory programs that infringe on their practice or the individual's right to freedom of choice. This allows them to accommodate members who oppose vaccination and promote harmful anti-vax theories to their patients. They are silent on the benefit to society from preventing the spread of infectious diseases.

Alternative medicine providers promote vaccine misinformation and conspiracy theories to sell pseudoscientific procedures, supplements, and oils to credible patients. Homeopathists, in particular, sell potions with little or no active ingredients that they allege have a natural vaccine-like effect. Many promote products that claim (without evidence) to cure the 'damage' caused by vaccines.

Anti-vaxxers seldom consider the public health consequences. One asymptomatic child with a highly transmissible disease like measles or COVID-19 can wreak havoc in a classroom, restaurant, or playground. The elderly, children too young to be vaccinated, and the immunocompromised risk severe illness and death from people who don't understand the science and fail to acknowledge the impact of their actions.

An unfortunate byproduct of the COVID-19 pandemic was that it expanded the anti-vax base into a political movement with the message that vaccines and vaccine mandates are an attack on personal freedom. Prominent anti-vaxxer Robert F. Kennedy Jr. has tilted the focus of his messaging away from alleged vaccine toxins like mercury and thimerosal to the idea that vaccines are a form of totalitarian oppression.

In mid-twentieth-century America, parents lived in fear of the polio virus. When Jonas Salk's polio vaccine was released in 1955, parents gratefully lined up their children for inoculations. Within two years, polio infections in the U.S. were reduced by over 90 percent. The World Health Organization started an initiative in 1988 to eradicate polio. With the help of Rotary International and the Bill Gates Foundation, this polio immunization program has been a fantastic success. In India, where the image of people crippled by the disease begging in the street was a familiar sight, polio was finally eradicated in 2014. Polio remains endemic (with just ten cases reported in 2023) in Afghanistan and Pakistan, and polio eradication is just around the corner.

Despite success stories for smallpox, polio, and COVID-19, we continue to see new vaccine-preventable disease outbreaks. Vaccination rates were falling in some parts of the world even before COVID-19 and fell further during the pandemic when parents stopped taking their children to the doctor for routine visits due to lockdown orders. Vaccination rates in children have yet to rebound even for routine vaccinations such as MMR (Measles, Mumps, and Rubella) and DPT (Diphtheria, Tetanus, and Pertussis).

Populist politicians and opportunistic celebrities cross the line from ignorant to evil when they step out of their area of expertise and promote anti-vax and anti-science nonsense. Opportunistic infectious diseases do not stop to consider your politics or your delusions.

2

ALTERNATIVE MEDICINE

THE TERM "SNAKE OIL" has become the catchall phrase to describe fraud in the health and wellness arena. Snake oil was initially brought to the United States by laborers from China who helped build America's railroads in the early 1800s. It was a real medicine, developed in antiquity from real snakes, that actually helped reduce inflammation. By the late 1800s, however, snake oil was marketed on the back pages of newspapers alongside other patent medicines, many of which contained alcohol (at best) or were actually harmful. They were soon lumped together as fraudulent cures collectively known as snake oil.

Snake oil salespeople today tout magnets, crystals, essential oils, and a range of therapies that have either not been proven to work or have been proven not to work. These therapies generally have a few things in common. First, they may convey some benefit, even beyond the placebo effect. Essential oils, for example, may provide pharmacological effects that would legitimately allow them to be classified as drugs if subjected to regulatory review. And they smell nice, which makes you feel good.

Second, alternative practitioners generally view themselves as holistic, treating the entire person rather than just the symptom. They wrongly dismiss traditional medical practitioners as focused only on the disease or drug, not the whole person. They also frequently combine therapies (acupuncture plus massage or spinal manipulation plus nutritional supplements, for example), obfuscating the true source of any positive outcome.

Finally, they rely primarily on anecdotal evidence to support their claims. They have not subjected their tools or techniques to the rigors of the regulatory review and approval processes. They are generally unable to explain the "mode of action" (exactly how it works in the body) for their techniques unless it is by using some "life force" that conveniently cannot be measured.

It is often difficult or impossible to tell from anecdotal cases whether a reported cure resulted from the treatment, a placebo effect, or the body's ability to heal itself. Was the anecdote unbiased and reported accurately? Was it fabricated? Can the results be repeated? Testimonial and anecdotal evidence sells, but it pales compared with the rigors of science and integrity reviews demanded by regulatory authorities such as the US Food and Drug Administration (FDA).

Many alternative medicine practitioners believe that human physiology and disease result from some unmeasurable, vital life force. They purport that chemical, electrical, and biological forces are separate and distinct from this life force and controlled by it. Many also believe that vast conspiracies exist to spread disease and suppress their practice.

Even the "true believer" practitioners understand that snake oil has pretty good gross margins and that health claims sell. They allow subjective personal experience to trump objective testing and intuition to supplant reason. They become not merely unscientific; they become anti-science. This is the heart of their delusion.

3

CHIROPRACTIC

The claim of alternative practitioners to not treat disease labels but the whole patient...allows alternative practitioners to live in a fool's paradise of quackery where they believe themselves to be protected from any challenges and demands for evidence.
—EDZARD ERNST

CHIROPRACTIC IS A FORM OF ALTERNATIVE MEDICINE based on the idea that most pain and disease result from misalignment of the spine. Chiropractors manipulate the spine and other joints and soft tissues to improve spinal alignment and treat a wide range of health problems.

Credit for the invention of this field goes to David Daniel Palmer back in 1895. Palmer was a strident opponent of traditional medical care (especially vaccines) and an avid spiritualist and magnetic healing practitioner before developing his big idea of chiropractic care. In fairness to Palmer, conventional medicine was not that great at the time.

Best practices in the late 1800's often included treatments that caused more harm than good, such as bloodletting with leeches or medicating with cocaine, opium, or alcohol.

Palmer was inspired to practice spinal manipulation by a ghost during a séance. He knew he was on to something when the realignment of a displaced vertebra cured deafness for one patient and provided relief from heart trouble for another. Those two data points convinced him that 95% of all diseases could be cured by chiropractic care and that there might be a business in all of this.

Palmer started a clinic and a school and wrote a textbook called "The Chiropractor's Adjuster, " published in 1910. His son, B. J. Palmer, is credited with helping mainstream the practice through promotion and recruitment. Because Palmer was anti-science, arguing against drugs, surgery, and even germ theory, the industry he founded still contains a significant population of anti-vax and anti-science practitioners who partner with acupuncturists, homeopathists, essential oil and nutritional supplement providers, and others in routinely making unsupportable health claims.

Samuel Homola, D.C. published "Inside Chiropractic—A Patient's Guide" in 1999 after a forty-year career researching and practicing Chiropractic medicine. He concluded that spinal manipulation could relieve certain specific back and neck pain and other musculoskeletal issues. Homola limited his practice to only those cases. He further concluded that no logical or scientific evidence exists that chiropractic can restore or maintain general health. Consistent with this conclusion, Medicare only covers manipulation of the spine, not other services or tests a chiropractor may order.

Dr. Edzard Ernst practiced in England as an academic physician, specializing in the study of complementary and alternative medicine. He is widely considered the world's first professor of complementary medicine, having started as an alternative medicine practitioner (homeopathy, acupuncture, herbalism, spinal manipulation, etc.) and then devoting the bulk of his career to studying the efficacy of those practices. He concluded that only five percent of alternative medicine is supported by evidence, and the rest either hasn't been validated by clinical research or has been proven not to work. He is also known as the "scourge of alternative medicine" for his courage in standing up for science and publishing scathing exposés about quack practices.

Quackwatch.org details numerous undercover assessments of chiropractic services. Diagnostic techniques range from X-rays to parlor tricks. In one instance, an Iowa undercover newspaper reporter was told by one chiropractor that his left leg was shorter than his right, and another said to him that his right leg was shorter than his left!

Chiropractors are quick to point to anecdotal evidence supporting spinal manipulation to relieve musculoskeletal and other ailments. Unfortunately, there is not enough confidence within the industry to subject these techniques to large-scale, double-blind clinical trials to determine specific therapies for specific indications. As a result, you will see a great deal of variation from one practitioner to another in both diagnostic techniques and the treatments for various conditions. The one thing most of these treatment plans have in common is that you need regular treatment to stave off disaster.

The National Institutes of Health (NIH) sponsors clinicaltrials.gov, a website containing a database of clinical trials. Search this database for ADHD, bed wetting, ear aches, or heart disease, and you will find very little evidence to support chiropractic care over traditional medicine. Even searches for clinical trials relating to chiropractic care of lower back pain (a chiropractic sweet spot) produce only a handful of studies, most using small samples (less than 100 patients) and showing little quantifiable benefit.

However, the NIH is pressuring the chiropractic industry to migrate to more evidence-based practices. NIH has partnered with Chiropractic colleges to fund clinical research to help determine whether the data supports chiropractic interventions. NIH grants require those institutions to pair with research-intensive medical schools to help improve chiropractic students' evidence-based practice skills. They are also encouraged to partner with physical therapists and physicians and use evidence-based outcome measures to quantify/determine results.

Early results from NIH-sponsored research support manipulative therapy for back, neck, and headache pain. Very little evidence supports chiropractic treatment for other indications. Unfortunately, the flotsam of validated scientific evidence supporting chiropractic care becomes lost in a sea of anecdotal evidence and pseudo-science.

Some people swear by their chiropractors. Traditional medicine did not solve their problem, and spinal manipulation did. But as long as the chiropractic industry refuses to embrace the rigors of the scientific method and regulatory approval, as long as they partner with quacks, and as long as they condone the anti-science and anti-vax branch of their community, they will remain relegated to the ignominy of magical thinking and delusion.

4

HOMEOPATHY

The plural of anecdote is not data.
—DR. RANDY FERRANCE

HOMEOPATHY WAS DEVELOPED IN THE 1790s by a German doctor named Samuel Hahnemann. It is based on two unconventional theories. The first is the "law of minimum dose," the idea that the lower the dose of a medication, the greater its effectiveness. In traditional pharmaceutical development, carefully controlled dosing studies are an important part of clinical trials. In homeopathic product development, active ingredients are diluted to the point where little or no measurable amount remains. If that seems counterintuitive, well, it is. It simply doesn't make any sense.

Second is a notion homeopathists call "like cures like," the idea that disease can be cured by substances that produce similar symptoms in healthy people. For example, onions make your eyes water and your nose run, so they are used in homeopathic remedies to treat hay fever. Treatments for other ailments may be made from poison ivy, arsenic, and other poisons. Online homeopathic supplier helios.co.uk even sells Tyrannosaurus Rex fossils diluted to the point of nonsense.

To make matters worse, even though the active ingredient that might be beneficial is diluted to oblivion, the inactive ingredients may contain substances that cause allergic reactions, have serious drug interactions, or are otherwise harmful or dangerous. Many liquid homeopathic products contain alcohol, and some contain heavy metals like mercury or iron. In 2017, the FDA discovered homeopathic teething tablets containing significant amounts of belladonna (deadly nightshade).

Homeopathy became quite popular in the United States in the late 1800's. By 1900, twenty-two homeopathic colleges and over 15,000 practitioners were operating in the United States. Traditional medicine was so ineffective at that time that patients of homeopathists often did better than those treated by conventional doctors. Homeopathy might not have helped, but at least it did not make things worse.

Even though the underlying concepts of homeopathy are entirely inconsistent with fundamental science, more than six million Americans use homeopathic products each year. Over $750 million is spent annually in the United States on products that, by design, have almost no active ingredients.

Pharmacy giants CVS, Walgreens, and Walmart undermine their credibility as healthcare providers by sporting racks of quack homeopathic medicines packaged to resemble conventional medication. CVS and Walmart were sued (unsuccessfully, unfortunately) by the Center for Inquiry for committing wide-scale consumer fraud and endangering their customers' health through packaging and product placement that made homeopathic medicines look like interchangeable matches with FDA-approved medicines.

Like most alternative medicine providers, homeopathists consider themselves holistic (treating the whole person, not just the disease). They also believe that physical illness has mental and emotional components. The homeopathic diagnostic technique includes assessing the patient's emotional and psychological state in addition to their physical symptoms. Remedies are then selected to treat the whole person, not just the disease. It is common in homeopathic practice for two people with identical physical conditions and symptoms to receive different remedies based on their mental and emotional states.

Patients who choose homeopathy over evidence-based medicine waste their money and are harmed by delaying or deferring effective and proven treatments. The use of homeopathy to treat serious infectious diseases is especially dangerous and delusional, not only for the patient but also for the community at large.

5

NATUROPATHY

Naturopathy is like going to a mechanic who fixes your car by burning sage and chanting.
—ANONYMOUS

NATUROPATHIC MEDICINE is a form of alternative medicine that lumps together a host of pseudoscientific treatment modalities under the broad rubric of natural medicine. Naturopathic techniques may include colonic irrigation, homeopathy, massage, acupuncture, electric current therapy, ultrasound, light therapy, or herbal medicine. They may also incorporate evidence-based techniques such as psychotherapy, exercise, or nutrition. They are all for it as long as it can be considered natural or non-invasive. Naturopaths typically advise against modern medical practices like vaccinations, drugs, and surgery.

Naturopathic medicine has its roots in late 19th-century Germany. It is a whole medical system emphasizing health restoration over disease treatment. Practitioners typically apply multiple techniques simultaneously to involve the mind, body, and spirit to facilitate the body's healing response.

It is difficult to generalize about naturopathic medicine because of the broad range of techniques that may be brought to bear by any given practitioner based on their training and scope of practice. They use so many different non-traditional methods of therapy that it is difficult to say precisely what the tenets of naturopathy are. Many present themselves as holistic primary care providers. More important than the physical exam is their assessment of the patient's diet, lifestyle, emotional state, and credulity. Their judgments on what is natural and non-invasive are suspect as well. Colonic irrigation (a common Naturopathic treatment method), for example, seems mighty invasive and just a little unnatural, especially in the absence of peer-reviewed clinical data to support its use.

While naturopaths aggressively lobby for professional recognition and laws to issue them medical licenses, regulation of the practice across the United States is highly variable. Some states license Naturopaths, while others don't regulate them at all. Eleven states permit Naturopaths to dispense drugs, and another eight states allow them to perform minor surgeries. Naturopathy is prohibited by law in South Carolina and Tennessee. While some jurisdictions allow Naturopaths to call themselves doctors, the lack of accreditation, scientific medical training, and measurable results means they lack the competency of actual medical doctors.

Naturopathic training amounts to a small fraction of training received by medical doctors practicing traditional medical care. Naturopathic practice is replete with pseudoscientific, ineffective, unethical, and potentially dangerous practices. Well-intentioned but seriously misguided advice can cause great harm and even death to vulnerable patients. And there is a pervasive culture of patient blaming among naturopathic practitioners. When something doesn't work, it's not because the therapy is ineffective but because the patient didn't do something right.

Even though Naturopathy may include some valid treatments and lifestyle advice, such as healthy sleep, a balanced diet, and regular exercise, it is rejected by the medical community overall because it relies on unproven and disproven alternative medical treatments. The least delusional course of action would be not letting them flush your colon or pocketbook.

6

ACUPUNCTURE

Even though I believe we should promote Chinese medicine, I personally do not believe in it. I do not take Chinese medicine.
—**CHAIRMAN MAO**

ACUPUNCTURE is an ancient Traditional Chinese Medicine (TCM) technique in which thin needles are inserted into the body at strategic pressure points to relieve pain, treat disease, or restore balance and harmony. The technique is thousands of years old, so acupuncturists promote its ancient roots, hoping to convince you that ancient practitioners discovered some medical magic that modern science does not understand.

According to TCM theory, good health requires alignment of qi (vital energy), which contains five elements: wood, water, fire, earth, and metal. Too much or too little of these elements disrupt vital energy.

TCM advocates believe that health is not merely the absence of disease but interconnectedness and the flow of life force. Acupuncture claims to open the blockage (or reduce the excess) of qi flowing through channels in the body called meridians.

Qi stubbornly resists any measurement, even using the most sensitive tools and techniques available to modern science. Pressure points (roughly 350) and meridians (14) are carefully plotted on diagrams of the human body. Unfortunately, they do not appear to have any physical structure that can be viewed or examined within the body. Despite numerous studies, the "mode of action" (how acupuncture functions physiologically) is undetermined.

Acupuncture is invasive, making it particularly difficult to construct blind control groups for clinical trials to prove efficacy. Testing is further confounded by variations in the size of the needles used, needle material, the number used, the length of time they remain inserted, the depth of insertion, the pressure points selected, and so on.

Some clinical testing has been done with control groups using non-penetrating needles or needles inserted at non-acupuncture points. This testing has determined that acupuncture is generally safe, but no solid evidence of benefit exists. While there may be a placebo effect, acupuncture is not scientifically supported as an effective healthcare method.

Despite the lack of scientific evidence, some anecdotal evidence supports the technique. Why do so many acupuncture patients report pain relief? The most likely explanation is the placebo effect, where patients experience a benefit from treatment even though the therapy is not accomplishing anything physiologically. The calming environment and the natural release of endorphins triggered by their belief are enough to convince many patients that their treatment is working. When traditional medicine has failed, some people are willing to try (and believe) anything that offers them hope.

In addition to the general reluctance of acupuncture practitioners to subject their techniques to the rigors of scientific testing and regulatory approval, they commonly partner with alternative medical therapies such as chiropractic, essential oils, and dietary supplements. Blending treatment methods makes it impossible to know what worked or didn't work and why.

Even though the procedure itself may do little harm, the delusion of acupuncture as a legitimate medical procedure can be harmful in several ways. Patients suffer needlessly when traditional medicine is delayed or abandoned in favor of alternative medicine, and their conditions may worsen. Acupuncture therapy can give people false hope and delay clinically proven and potentially more effective treatment.

7

ESSENTIAL OILS

*Always carry a flagon of whiskey in case of snakebite
and furthermore, always carry a small snake.*
—W.C. FIELDS

ESSENTIAL OILS ARE MADE by pressing seeds, flowers, bark, leaves, fruit, or other components into a concentrated oil to capture "essential" aromatic compounds. The fundamental problem with essential oils is that they are promoted to treat health conditions based primarily on anecdotal evidence. They are not developed, tested, regulated, or dispensed as drugs, yet they are sold as drugs with no science or regulatory approval to support their claims.

The United States Food and Drug Administration (FDA) regulates and approves drugs based on claims for intended use. Applications for approval must explain the "Mode of Action" (exactly how it works in the body), demonstrate safety and efficacy, and document the risk of side effects and contraindications for each intended use or health claim. This is generally done using a series of progressively larger and more comprehensive clinical trials with statistically significant sample sizes of affected populations, frequently using double-blind, randomized comparisons with placebo groups or current standard-of-care groups. The application is then submitted to a panel of experts to assess the data and determine if it is statistically safe and effective enough (compared with the placebo group or existing therapies) to warrant approval. The FDA will then consider the advisory panel recommendation and decide whether to approve it.

As you may imagine, this is tedious and expensive, so essential oil producers and marketers have shied away from the regulatory approval process. Instead, their attorneys have cynically crafted language that dances right up to the line of making a disease claim without stepping over it. Here are a few examples from one major producer's website:

DISEASE CLAIM	NOT A DISEASE CLAIM
Fights cold and flu	Supports the immune system
For ear infections	For occasional ear discomfort
Relieves insomnia	For occasional sleeplessness
Relieves headaches	Relieves head and neck tension
Helps fight obesity	Helps with weight loss plan
Prevents osteoporosis	Supports bone health
Fights germs, viruses, bacteria, or allergens	For seasonal threats

They also routinely disclaim their product labeling and marketing materials in small print with phrases like:

"These statements have not been evaluated by the Food and Drug Administration."

and

"This product is not intended to diagnose, treat, cure, or prevent disease."

The largest suppliers of essential oils (companies like doTERRA and Young Living) sell their products through independent representatives. These independent representatives are significantly less cautious and routinely make health claims they are not legally allowed to make.

In May 2022, the FDA reviewed websites and the social media channels of various Young Living independent representatives (called Brand Partners) and found that they were making claims that the FDA could only interpret as drug claims. This is the first time the FDA has gone after independent representatives, issuing warning letters demanding that they remove health claims or face seizures, fines, or even jail time. Young Living forced their independent sales force to clean up their act, at least on public marketing platforms. It remains to be seen if this will extend to in-person sales and marketing.

Essential oils can also be found in alternative health environments (such as chiropractic and acupuncture clinics), networking groups, and exercise and yoga studios. Health claims made verbally at multi-level marketing parties, or other in-person sales environments, are very difficult for the FDA to regulate or control.

Essential oil marketers routinely declare their products pure and natural, implying that this somehow prevents them from being harmful. Some even go so far as to claim they are free of toxic chemicals, not recognizing that the harm of any chemical is in the dosage. The most delusional (or at least unthoughtful) claim is that they are chemical-free, which is absurd.

If you like how essential oils smell or make you feel, that is great. However, you can easily dismiss as wishful thinking any claims of health benefits not backed up by peer-reviewed science and approved by the FDA, no matter how compelling the testimonials seem.

8

DIETARY SUPPLEMENTS

*Health nuts are going to feel stupid someday,
lying in hospitals dying of nothing.*
—REDD FOXX

DIETARY SUPPLEMENTS are manufactured products intended to supplement one's diet to achieve some health benefits. This includes products ranging from herbs to vitamins to probiotics. The dietary supplement industry is estimated to generate over $150 billion annually, with over 50,000 products marketed in the United States alone. Dietary supplement ingredients may be extracts or concentrates, including vitamins, minerals, herbs, botanicals, etc.

The U.S. Food and Drug Administration (FDA) regulates dietary supplements as food products, meaning they only monitor for accuracy in advertising and labeling. The U.S. National Institutes of Health (NIH) plainly states that dietary supplements "are not medicines and are not intended to treat, diagnose, mitigate, prevent, or cure diseases." Since nutritional supplements are minimally regulated, they are usually sold by marketing hype rather than clinical evidence. As such, the industry attracts more than its share of quacks and flakes, especially in the multi-level marketing arena.

The most common dietary supplement is multivitamins. The NIH states that multivitamins "may be of value" for those who are nutrient-deficient in their diet. Multivitamin marketers accept that faint praise as a full-throated endorsement. They market their products in health stores and through multi-level marketing programs where store clerks and independent sales representatives can make unsupported health claims with impunity.

Another common dietary supplement is herbs. Herbs may have some nutritional or medical value, but the U.S. Congress lumped them in with other dietary supplements in the 1994 Dietary Supplement Health and Education Act. Unless herbal product producers make explicit medical claims in their packaging or marketing materials, their products are regulated as food, not drugs. As a result, labeling and processing standards are not very rigorous. Contents and potency are often not disclosed or disclosed incorrectly on the label. Market studies have found that the ingredients and doses of herbal products may vary considerably from brand to brand and even from batch to batch.

One of the core beliefs of herbalists (and nutritional supplement marketers in general) is that natural and organic herbs are superior to synthetic compounds. They argue that herbs are the source of many modern medicines. This is true, but some important facts are omitted. Drug products are studied and refined to isolate stable, potent, and predictable active ingredients for more precise dosage control. So-called natural herbal remedies can vary significantly from batch to batch and may contain other ingredients that have not been researched and may cause harm.

Many people believe that organic products (those grown without pesticides or synthetic fertilizers) are superior in some way to those produced by conventional agriculture. Understandably, some would be willing to pay more for lower-yield products made in this manner. The notion that inorganic nutrients such as nitrogen, phosphorus, and potassium from the soil are somehow better than those from a factory demonstrates a complete misunderstanding of chemistry, plant nutrition, and physiology. Plants have no mechanism for distinguishing where their nitrogen comes from. It is there or not, and the plant responds accordingly.

Some supplement providers have cynically started calling their products "nutraceuticals." This clever bit of wordplay implies pharmacologic effects for products they know they cannot legally promote as drugs. One prominent "nutraceutical" marketer makes a dramatic show of absorbability. Their multi-level marketing representatives will hold up a small plastic cup, which they will compare to the size of the opening in a cell wall. They will show how a basketball or baseball cannot fit into the cup, but a golf ball can. They claim their products are refined to be small enough to be absorbed through the narrow cell wall opening, implying that other products are not.

They would have you believe that they alone have mastered the concept of absorbability that the rest of the world has missed. Their marketing (basketball versus golf ball) attempts to trump the comprehensive scientific understanding of how a molecule moves through the body known as ADME (Absorption, Distribution, Metabolism, and Elimination), as required in the clinical research and development for FDA-approved products.

Another marketing channel for dietary supplements is retail health-food and nutrition stores. Most retail clerks at these stores have little or no formal nutrition or health care training. Yet they routinely diagnose ailments and prescribe products from their stores. This is illegal (practicing medicine without a license) and frighteningly dangerous.

The FDA maintains a health fraud database listing over 2,000 unapproved products that the FDA has called out in warning letters, recalls, and press releases for falsely claiming to cure, mitigate, treat, or prevent disease. The FDA admits that this is only a tiny fraction of the potentially hazardous products marketed to consumers online and in retail establishments. The supplement industry is rife with risk, fraud, and misinformation. It is better to eat a balanced diet and avoid the delusion of dietary supplements altogether.

9

DETOX SCHEMES

*As for those grapefruit and buttermilk diets,
I'll take roast chicken and dumplings.*
—HATTIE MCDANIEL

TRADITIONAL MEDICINE uses the word "detoxification" for therapies to minimize withdrawal symptoms for drug and alcohol abusers. Alternative medicine uses the term to raise concerns about the danger of accumulated toxins in the human body.

The notion that we may accumulate and store toxins in the body over time seems logical enough. Toxins are everywhere. Mercury is in fish, lead is in old paint and toys, smog is in the air, pollutants are in the river, and forever chemicals are in the water supply. If toxins can get into fish, they can undoubtedly get into us.

Alternative practitioners play to those fears. However, there are a couple of problems with the whole notion of detox. The first is that with any substance, the dose makes the poison. Many toxins accumulate in such tiny amounts that they are insignificant to our health and well-being.

Not long ago, there was a scare about formaldehyde in fruit juice. Formaldehyde, in large doses, can cause cancer and other frightening diseases. Formaldehyde is a naturally occurring substance in apples, bananas, grapes, and many different fruits and vegetables. But the parts per million of formaldehyde found in juice is inconsequential. And you know what else? According to the European Food Safety Authority, "Formaldehyde is an important metabolic intermediate that is physiologically present in all cells." In other words, it is already there! But what if you drank a whole barrel of juice and got too much formaldehyde? Well, formaldehyde is water-soluble and has a half-life of about one minute. This means that 99.9 percent of the formaldehyde from your barrel of juice would be gone within minutes of drinking it. Consuming a whole barrel of juice would kill you before the formaldehyde ever had a chance.

More importantly, the human body has highly effective natural detoxification mechanisms. The liver filters blood in the body and is really good at breaking down and flushing out poisonous substances like drugs or alcohol. Kidneys cleanse the blood of toxins and transform waste into urine. Our digestive tract dissolves and purges pathogens and other foreign substances, and the skin releases toxins via sweat.

Detox proponents are generally unable to identify the specific toxins to be eliminated by a given treatment. But toxins are not nebulous generalities. They have names like mercury, lead, arsenic, sulfur dioxide, or DDT. If a detox product producer wanted to claim that their product eliminated some specific toxin, it would be easy enough to conduct an independent study to confirm this. Set up and measure initial toxin levels in two groups of people; then, half get your product, and the other half get a placebo. Measure again in thirty days and publish your results in a peer-reviewed medical journal. Voila! Proof. Comprehensive clinical testing is not done, however, because the resulting data is unlikely to support the detox claim.

The most widespread detox scam is fruit or vegetable juicing (also called cleansing). This detox diet involves consuming only fresh fruit or vegetable juices for three to seven days. Juicing detox advocates promise that the juice cleanses will flush out toxins from the body, help you lose weight, and improve your overall health. Unfortunately, juicing is not healthier than eating solid foods. It is probably worse for you in terms of overall health. Liquids lack fiber or protein. Without fiber, you get hungry faster. Since juice is often mostly sugar, a juice cleanse diet throws away the best things about fruit and keeps the worst. Most people lose weight on a juice detox, mainly by consuming dangerously low levels of calories during the cleanse. Any other low-calorie weight loss diet would yield the same results.

Another detox scam made famous by celebrity endorsement is the Master Cleanse Detox Diet. Dieters drink only water, lemon juice, maple syrup, and cayenne pepper for ten days to two weeks. This scheme has many variations, some adding salt water or herbal teas. Master Cleanse Detox kits are sold online for $85 to $125 each.

The Master Cleanse Detox is an unhealthy way to lose weight. It is deficient in calories and essential nutrients such as protein, carbohydrates, fats, fiber, vitamins, and minerals. People often suffer headaches, dizziness, diarrhea, nausea, and exhaustion while undergoing this cleanse. Many people will lose weight on this diet due to deficient caloric intake. However, most people also regain weight quickly when they resume eating normally.

Another pseudo-scientific detox therapy is called chelation. The FDA has actually approved Chelation therapy to treat extreme cases of metal poisoning. Chelation drugs bind to metals in your blood so your body can remove them through urination. Alternative medicine chelation therapists use this drug and the subsequent urine testing as a parlor trick to convince you that you are suffering from poisonous metals in your blood.

Chelation therapy scams can be dangerous. Chelating drugs can remove minerals like calcium, copper, and zinc that your body needs. Low calcium levels in the blood can lead to kidney damage.

Another category of detox quackery is the foot pad or foot bath scam. Patients are told to wear foot pads overnight while they sleep or soak their feet in low-voltage water to "suck out the toxins." Practitioners cite discolored water or foot pads as evidence that harmful toxins have been removed. Laboratory testing of discolored foot pads and water determined that no toxins were collected and that the discoloration occurred due to chemical reactions that were easily repeatable without the use of feet.

Another detox-related scam is colon cleansing. Colon cleansers are typically powders consisting mainly of fiber (such as psyllium or flaxseed) and laxatives (such as cascara or magnesium oxide). They may also contain probiotics, vitamins, minerals, herbal teas, etc. One example that gained traction in the 1990's was coffee enemas as a quack treatment for pancreatic cancer.

Even though the FDA may have approved some of the active ingredients in these products, it has not approved them for a colon cleansing indication. Colon cleansing proponents falsely claim they are removing toxins and treating other health disorders. Any physician who has performed an intestinal exam will attest that the theory of toxin buildup in the colon is nonsense. Colon cleansing products confer no health benefits and can cause unnecessary bloating, cramps, and diarrhea. The Journal of Clinical Gastroenterology described colon cleansing as not merely useless but potentially dangerous.

All detox therapies are risky. Potential side effects include liver damage, malnutrition, anemia, muscle loss, heart palpitations, and a weakened immune system. Most detox diets lack protein, a critical nutrient for detoxifying the body naturally using the liver and kidneys. They also lack the dietary fiber your digestive system needs to process waste properly. The green tea extracts of many detox diets have been associated with an increased risk of liver injury.

DETOX SCHEMES

Any use of the words detox, cleanse, or toxins by alternative medicine providers should be viewed with suspicion. Suppose you want to help your body avoid toxic substances. In that case, there are some straightforward steps to take: minimize your environmental exposure, exercise to the point where you sweat, fast occasionally, get quality sleep, and drink lots and lots of water. Don't let the detox delusion cleanse your wallet.

10

CUPPING

Cupping is a form of quackery with no scientific basis.
—THE LANCET

CUPPING THERAPY involves placing glass, plastic, bamboo, or ceramic cups on the skin and then pumping air out of the cup to create suction. Cupping can trace its roots to ancient Egyptian, Chinese, and Middle Eastern cultures. The theory is that suction creates a negative pressure environment to lift muscle fibers and draw blood to the area. Cupping practitioners claim it improves qi (vital energy or life force), removes toxins (which remain unspecified), and improves blood flow for pain relief and musculoskeletal injuries like strains, sprains, and back injuries.

Cupping is practiced using a wide array of techniques. Stagnant cupping involves leaving the cup in place while creating a negative pressure environment for five to fifteen minutes, often resulting in a circular hickey-like mark on the skin. Dynamic cupping involves sliding the cup across the skin. Wet cupping goes a step further, making a tiny cut on that suctioned area of the skin and then using a second suction to draw out a small amount of blood. Fire cupping involves using a flame to heat the cup to create suction as it cools.

Most of the research on cupping has been of low quality due to bias, small sample size, and failure to control critical variables associated with the practice. Most of the research focuses on how much cupping techniques reduce pain. Few seek to understand how it works. Those who believe cupping is good for pain relief or enhances sports performance cannot demonstrate if this is due to the placebo effect, the effect of relaxation in a caring environment, or the cupping itself.

Ultimately, there is no evidence that cupping effectively treats any disease or health condition. Notions of vital energy and toxin removal are the standard fare of pseudoscientific quackery. The claim that cupping improves blood flow is absurd on its face. The hickey or bruise created by cupping is a blood clot, the opposite of improved blood flow.

Cupping has gained popularity with endorsements from sports celebrities like NFL player DeMarcus Ware and Olympic swimmer Michael Phelps. Phelps was observed in the 2016 Olympics with multiple purple bruises on his back resulting from cupping. Phelps apparently believes cupping helps speed up recovery. Images of Phelps' Olympic performance with cupping bruises are standard fare on cupping websites worldwide. Unfortunately, even world-class athletic ability cannot protect you from the appeal and delusion of pseudoscience.

11

EAR CANDLING

He was not so much brain as earwax.
—WILLIAM SHAKESPEARE

EAR CANDLING (also known as coning) involves placing a cone-shaped device in the ear canal to extract earwax and other impurities using smoke or a burning wick. The procedure allegedly creates a low-level vacuum that draws wax and other debris out of the ear canal.

Proponents claim that candling can relieve sinus pressure, cleanse the ear canal, improve hearing, purify the mind, cure ear infections, stabilize emotions, purify the blood, cure cancer, release blocked energy, align the chakras, etc. This wild mix of unsubstantiated medical claims and new-age doublespeak is both delusional and dangerous.

Ear candling does not remove earwax, toxins, or anything else from the ear canal. Practitioners collect the residue of the procedure in a bowl to show all the toxins that have been removed from the ear canal, sinuses, and brain, even though basic anatomy shows that the ear canal is not connected to any structures beyond the eardrum. Repeated demonstrations have shown that candling, with or without an ear nearby, produces precisely the same amount of toxic-looking gunk.

Ear candles have been associated with injuries such as skin/hair burns and middle ear damage, such as punctured ear drums. In 2007, the FDA issued an alert identifying ear candles as dangerous when used as recommended by the manufacturers. Ear candles cannot legally be imported into or sold in Canada because Health Canada has determined that there are no other reasonable uses other than medical, and any medical claims have not been substantiated.

Despite all this, ear candles are still widely available on the Internet and at health-food stores. They are also an ancillary service offered at many alternative medicine practices. Delusions of a feather flock together.

12

REIKI

The next big frontier in medicine is Energy Medicine.
—DR. MEHMET OZ

REIKI is a Japanese alternative medicine tradition of energy healing where "universal energy" is transferred through the hands of the practitioner to the patient to provide emotional or physical healing. Reiki comes from a combination of Japanese words for soul and vital energy.

Reiki is similar to (and has largely replaced) Therapeutic Touch, which was popular in the United States in the 1980s and 1990s. Therapeutic Touch was discredited in 1996 when a nine-year-old budding junior scientist submitted a clinical trial of the practice for her fourth-grade science project. Twenty-one Therapeutic Touch practitioners were subjected to blind tests and failed to produce anything besides random results. The results were such a compelling dismissal of the practice that they were later replicated and published in a legitimate medical journal. Therapeutic Touch faded from prominence, and Reiki took its place.

Like many other types of alternative medicine, Reiki is a form of vitalism, a pre-scientific belief that relies on the premise of a universal life force that separates humanity from non-living things. This life force (often called qi) attempts to explain everything we don't understand about how the body works. In ancient medicine, this served as a convenient intellectual placeholder. But as modern science has helped us to understand how the body works at the chemical and cellular level, the notion of vital life force is more or less running out of things to be responsible for.

Vital life force energy is impervious to measurement, placing it squarely in the realm of delusion. Attempts to transfer or manipulate this energy for therapeutic effect are doomed to fail and ethically questionable. Reiki is harmful to the extent that it delays or preempts treatment with legitimate medical therapies that work. At a minimum, Reiki is a deception (even if the practitioner embraces the delusion) that encourages patients to rely on magical thinking instead of clinically proven treatments.

13

ALTERNATIVE MEDICINE FOR PETS

> *Animals are such agreeable friends—*
> *they ask no questions; they pass no criticisms.*
> —GEORGE ELIOT

FOR THOSE WILLING TO DOUBLE DOWN on their delusions, alternative medicine can also be applied to animals, with the added benefit of the patient's inability to communicate about symptoms or results.

Established veterinarians who incorporate alternative medicine into their practice generally do so because their customers ask for it and because it is an additional source of revenue. Many treat their alternative medicine offerings as complementary to their traditional services, making distinguishing the source of any positive results impossible.

However, some veterinarians perform only alternative medicine and may even specialize in one specific type of alternative medicine. There are enough of these quack specialists that they have formed professional associations such as the International Veterinary Acupuncture Society, the Veterinary Medical Aromatherapy Association, and even the American Veterinary Chiropractic Association. The American College of Veterinary Botanical Medicine promotes the use of medicinal plants, and the Academy of Veterinary Homeopathy is dedicated to understanding and preserving the principles of classical homeopathy for pets.

These so-called holistic veterinarians seem to believe in what they do despite the lack of scientific evidence that the tools of their trade prevent or cure any illness in pets. Their websites are awash in testimonials. Links to clinical studies and peer-reviewed science are conspicuous by their absence. Lacking evidence, holistic practitioners judge their results using unquantifiable measures such as appetite improvements or how a dog wags its tail and carries its ears. One certified homeopathic veterinarian said, "We don't care how it works. The bottom line is, you can be a successful homeopath without knowing how it works."

While many of these alternative medicine practices are not harmful, some are. Sometimes, they have adverse side effects or interfere with other medications. Some can even be fatal. Herbal remedies and essential oils, in particular, can be poisonous to pets. Holistic veterinary practitioners who prescribe untested compounds (Yes, but they are natural!) based on gut feel and anecdotal evidence are particularly cruel and irresponsible.

Subjecting animals to sham treatments also harms them. Animals given homeopathic remedies, for example, have no option but to continue to suffer, no matter how much their ears perk up or how much their owners believe in homeopathy. There can be no placebo effect if you don't understand that the ersatz medicine you were given is supposed to make you feel better.

Further along the delusion scale, energy healers use chakras and crystals to elevate their dogs to higher states of consciousness and release traumatic puppyhood memories. Social media platforms like TikTok are awash with self-described intuitives and animal communicators. Veterinary Reiki practitioners are happy to take your money to align your pet's qi. One compassionate yet enterprising Reiki healer will even work their magic over the internet! Just send in your $35 and wait for the waves of healing energy to wash over your pet. They could run the palms of their hands over your wallet to equal effect.

If you love your pets, leave their chakras alone.

14

FAD DIETS

My doctor told me to stop having intimate dinners for four. Unless there are three other people.
—ORSON WELLES

THERE ARE TWO SIMPLE TRUTHS about dieting. First, you will lose weight if you burn more calories than you consume. Second, most diets fail, especially over time. The real challenge of dieting is doing so in a healthy and sustainable way.

Fad diets are often based on incomplete or faulty research and are promoted as quick and easy weight loss secrets. They may involve banning entire food groups, such as fats or carbohydrates. Or they will hype the benefits of a particular "superfood," such as acai berries, grapefruit, or cabbage soup. They may also offer techniques for consuming fewer calories, such as intermittent fasting.

Fad diets can leave you at risk of missing out on essential nutrition, and their effects are usually not sustainable. Any diet restricting the number of calories consumed will result in initial weight loss. By one estimate, most dieters only stick to their diets for four to six weeks. Once the diet stops, most people quickly regain weight. One UCLA study found that after five years, 50% of dieters exceeded their initial weight by 11 pounds.

Fad diets rarely focus on lifestyle modifications like exercise and proper sleep to keep weight off. Highly restrictive diets leave dieters feeling deprived and lead to cravings that lead people to cheat. Most eventually give up.

Inventors and promoters of fad diets generally have something to sell. The Bulletproof Diet comes with ten basic principles, the tenth being that you need to buy their supplements for nutritional support. A naturopath invented the Blood Type Diet with a book containing lots of scientific mumbo jumbo but without the benefit of any actual science to support their conclusions. The American Journal of Clinical Nutrition (a peer-reviewed biomedical journal) said, "No evidence currently exists to validate the purported health benefits of blood type diets."

Some fad diets are harmful. The Grapefruit Diet can trigger adverse events for people using statin drugs like Lipitor, or transplant anti-rejection drugs like Neoral and Sandimmune. The Cabbage Soup and the Master Cleanse Diet are dangerously low in calories and can cause nutrient deficiencies and headaches. The Atkins Diet and the South Park Diet are very high in fat and can raise cholesterol levels and increase the risk of heart disease. The Cotton Ball Diet (eating cotton balls soaked with juice to help you feel full) is just plain nuts, as cotton balls are not digestible and can be toxic or cause intestinal blockages.

Most fad diets (like most religions) have a book. They also have websites hawking subscriptions, meal plans, recipes, coaching, supplements, etc. Fad diets can be expensive, especially if you have to buy special foods or supplements. They are not balanced or sustainable (eliminating whole food groups or relying on a supposed superfood) and are not comprehensive in their approach (focusing only on calories in, not calories out).

The delusion of fad diets is that they are based on unrealistic and unsustainable principles supported by dubious testimonials and celebrity endorsements.

15

BIODYNAMICS

The health of soil, plants, animals, and man is one and indivisible.
—SIR ALBERT HOWARD

THE BIODYNAMIC ASSOCIATION describes Biodynamics as "a holistic, ecological, and ethical approach to farming, gardening, food, and nutrition." Biodynamics is similar to other types of organic farming in that compost and manure are encouraged, while synthetic (not derived from natural sources) fertilizers, pesticides, and herbicides are forbidden.

Biodynamics goes beyond ordinary organic practices by treating animals, crops, and soil as a single system and using an astrological sowing and planting calendar. It also relies on herbal and mineral additives prepared using methods that are more magical than farming.

Biodynamic compost is enhanced using medicinal herbs from yarrow, chamomile, stinging nettle, oak bark, dandelion, and valerian. Practitioners argue that this enriched compost serves many purposes, including bringing more sensitivity to the composting process, helping attune the soil to the whole farm organism, and restoring balance to the climate. Biodynamic sprays made from horsetail tea, cow manure, and quartz crystals (buried inside a cow horn at the proper phase of the moon) help bring plants into a dynamic relationship with soil, water, air, warmth, and cosmos, plus other New Age word salad.

Biodynamic agriculture was developed in 1924 by Dr. Rudolf Steiner, an early organic farming advocate. It has great appeal in its approach to biodiversity and its view of the farm's plant, animal, and other elements as an integrated system. However, it goes off the rails when it relies on spiritual, mystical, and astrological pseudoscience to guide when to sow, cultivate, harvest, or bury cow horns.

Biodynamics is practiced in more than 50 countries worldwide, particularly in viticulture. More than 600 certified biodynamic vineyards exist worldwide, nearly half of which are in France. Many others follow biodynamic rules without enduring the cost and rigors of certification.

One aspect of Biodynamics that has caught on in the wine world is using astrology and the lunar calendar to determine the most auspicious days to plant, compost, harvest, and even drink and taste wine. Under this theory, the lunar calendar is divided into Root, Flower, Leaf, and Fruit days.

- Fruit days are when the moon is in the Fire signs of Aries, Leo, and Sagittarius. They are the best days to harvest grapes and drink wine.

- Flower days are when the moon is in the Air signs of Gemini, Libra, and Aquarius. The vineyard should be left alone on Flower days. Spend that time drinking aromatic wines such as Viognier or Torrontes.

- Root days are when the moon is in the Earth signs of Capricorn, Taurus, or Virgo. Root days are best for pruning, and bad days for tasting.

- Leaf days are when the moon is in the Water signs of Cancer, Scorpio, and Pisces. Leaf days are good for watering and bad for drinking or tasting wine.

So, if you like a particular wine on a Fruit Day, you might not like it on a Leaf Day. And if you don't like it on a Root Day, it may not be your fault. Check your lunar calendar and try again.

16

MONSANTO AND THE GMO DELUSION

All the food we eat—every grain of rice and kernel of corn—has been genetically modified. None of it was here before mankind learned to cultivate crops. The question isn't whether our food has been modified, but how.
—MICHAEL SPECTER

GMO STANDS FOR GENETICALLY MODIFIED ORGANISM. While humans have genetically modified plants using cross-pollination and selective breeding for thousands of years, the science of genetic engineering has only existed for about 50 years.

Genetic engineering involves copying a gene with a desired trait from one organism into another or editing a gene to provide that trait. Genetic engineering continues to find new applications based on advances in the study of plant and animal genetics. The term GMO is commonly used to describe genetically engineered food.

Most people who oppose genetically modified foods do not oppose genetic modification by cross-pollination or selecting and propagating seeds with desirable traits. It seems to be the application of science that crosses the line from "natural and good" to unnatural and harmful to human health and the environment. Many GMO opponents are particularly horrified by gene editing, especially transferring genes across species, not realizing that humans share more than 50 percent of their genetic information with plants and animals and that the integration of genetic material across species occurs frequently in nature and not just in laboratories.

Genetic engineering has grown in use because it is fast and precise. Scientists edit plant and animal genes to improve yield, increase insect and virus resistance, improve drought and herbicide tolerance, and enhance flavor, appearance, and shelf life. A 21-year study of GMO corn yield data published in Scientific Reports found that yields increased as much as 24 percent over non-GMO corn and dramatically decreased dangerous food contaminants.

The best-known example of a genetically modified food may be Golden Rice. Genetic engineering was used to alter Golden Rice to produce beta-carotene, which is not ordinarily present otherwise. Golden Rice is now being grown in areas with a shortage of dietary vitamin A. Vitamin A deficiency causes irreversible blindness and death for hundreds of thousands of children under five each year. Golden Rice gained its first approvals for use as food in 2018 despite vigorous and sustained resistance from anti-GMO advocates. Today, Golden Rice saves lives and prevents blindness for thousands of children each year in places like the Philippines and Bangladesh.

Genetic engineering has potential downsides, however. Early successes have tended to lessen biodiversity and can lead to monoculture. GMO plants can overtake or crowd out native species. Opponents of genetic engineering have made claims about increases in cancer, immune system responses, infertility, congenital disabilities, and even changes in human DNA. Even though these claims are largely unsubstantiated, there are risks that we need to understand better.

The poster child for GMO food and genetic engineering was the Monsanto Company (acquired by Bayer in 2018). Monsanto was a Fortune 500 biotechnology and agribusiness concern best known for creating the herbicide Roundup and a broad range of genetically engineered seed products.

Moralistic notions of food sanctity, fears of contamination, and Monsanto's litigious approach to protecting their intellectual property inspired an intense emotional response among GMO opponents, who concluded that Monsanto was the devil and paraded their outrage in public crusades against the company. The March Against Monsanto movement was formed in 2013 in response to a failed California ballot initiative on GMO food labeling. At its peak, this group held hundreds of protests and marches worldwide. Since Bayer acquired Monsanto in 2018, the organization has faded from prominence. Its Facebook page has over a million likes but has become a cesspool of anti-vax and conspiracy theory nonsense.

Ultimately, the GMO delusion is the idea that genetic changes involving test tubes are somehow unnatural and immoral. GMO opponents would allow their fear and misunderstanding of the science to prevent improvements in human health, food safety, and production.

VOLUME 2

YOUR MONEY (FINANCIAL DELUSIONS)

When I was young, I thought that money was the most important thing in life; now that I am old, I know that it is.
—OSCAR WILDE

INTRODUCTION

Financial delusions are beliefs about money or assets not supported by rational valuations. A bubble, for example, is characterized by rapidly increasing prices for an asset followed by an even more rapid crash in value. In other words, the bubble inflates and then suddenly pops.

Charles Mackay wrote about 19th-century asset bubbles like the Mississippi Scheme and the South Sea bubble that were too good to be true until they weren't. He also beautifully detailed the 17th-century Dutch Tulip Bulb mania, which captured the imagination (and the money) of rich and poor alike. Delusions like these are not just in the distant past or something that only happens far away. They can be as close as your social media account.

Former Federal Reserve Bank Chair Alan Greenspan coined the phrase "irrational exuberance" to describe market behavior during the price increase phase of a bubble. Investors feel really smart watching their asset values grow, and continued price increases reinforce their delusion. Few realize they are in a bubble until it is too late.

17

DOT-COM BUBBLE

Unlike with other famous bubbles ... the Internet bubble is riding on rock-solid fundamentals, perhaps stronger than any the market has seen before.
—HENRY BLODGET

WEB BROWSERS ARE UBIQUITOUS TODAY, but when the Netscape web browser was introduced in 1994, it was revolutionary. Before Netscape, internet-based communication had some traction in the military, industry, and higher education. However, the tools available to access the Internet were clumsy and technical. Netscape provided the first easy-to-use, mass-market tool for browsing the World Wide Web. By 1995, over 40 million people were online using dial-up modems to access the Internet over Plain Old Telephone Service (POTS lines).

Netscape's Initial Public Offering (IPO) followed in 1995, and stock initially offered at $28 per share closed the first day at $58.25. This gave a start-up company with no profits a market valuation of $2.9 billion. This heady success became known as a "Netscape Moment." Everyone wanted a Netscape Moment.

The formula for achieving a Netscape Moment was to raise venture capital with an idea and a catchy domain name like etoys.com, webmd.com, or hothothot.com. Get big fast, and if your concept is good, growth will follow. More importantly, an IPO will result in a huge payoff for the founders, and eventually, maybe—profits will follow.

Bidding wars were launched, and lawsuits were filed over domain names like MTV, RoadRunner, and McDonald's. People for the Ethical Treatment of Animals (PETA) was horrified to learn that a gentleman named Michael Doughney owned a website for People Eating Tasty Animals (peta.com). PETA sued and eventually won its domain name back under the Anti-Cybersquatting Consumer Protection Act of 1999.

Low interest rates and a lower marginal capital gains rate from the Taxpayer Relief Act of 1997 contributed to the bubble by making venture capital easier to raise. Investors were eager to invest in any dot-com company regardless of the fundamentals. Traditional financial metrics were abandoned for companies with growth potential, with or without earnings. It was a new world, and the old rules didn't seem to apply anymore.

The technology-heavy Nasdaq Composite stock market index spent most of 1994 under 400 but increased 85% in 1999. It peaked in March of 2000 at over 4,600, more than a ten-fold increase. In March 2000, the Price/Earnings Ratio (P/E Ratio) for the Nasdaq reached a mind-blowing 175, compared with a stock market historical average of 13 to 15 and a technology company average of 20 to 25. Individual investors quit their day jobs to trade on the financial market, finding that everything they touched turned to gold. Dot-com founders and employees became instant millionaires on paper.

Many dot-com companies incurred hefty operating losses by offering their services or products for free or at a discount while spending lavishly on advertising and promotions. Mind share and brand awareness were crucial for expected future profits. Seventeen dot-com companies bought ad spots for Super Bowl XXXIV in January 2000 at roughly $2 million for each 30-second spot. Two of those (Pets.com and Epidemic.com) were both defunct by the end of that same year.

THE BUBBLE BURSTS

All this exuberance seemed like the natural state of things to those immersed in it. But it only takes a tiny pinprick to burst a bubble, and by March of 2000, many sharp pins were reaching for that bubble. That month, the Federal Reserve raised interest rates in an ongoing effort to slow the economy and prevent inflation; Japan had again entered a recession, triggering a global technology stock sell-off; Barron's Magazine featured a cover article titled "Burning Up; Warning: Internet companies are running out of cash—fast." Bloomberg News suggested it was finally time to pay attention to the numbers.

MicroStrategy (a leading software and services company) announced a revenue restatement that caused its stock price to fall 62% in one day. Microsoft was found guilty of violating the Sherman Antitrust Act, leading to a one-day 15% decline in the value of Microsoft shares and a 350-point (8%) drop in the Nasdaq. The week ending April 14, 2000, saw the Nasdaq Composite index fall 25%, as many investors were forced to sell stocks ahead of Tax Day.

This sobering news led many people to rethink their investments. By the end of 2000, many Internet stocks had declined in value by 75% from their highs. The Nasdaq Composite Index fell to 2,300, half the peak in March, erasing over $1.75 trillion in market value.

One of the more spectacular failures was Pets.com. With Amazon as a significant investor, online pet store Pets.com raised $82.5 million in a February 2000 IPO. Its business model was riddled with holes, however. It couldn't economically ship bulky items like bags of pet food. It had competition from local retailers and other online pet supply start-ups. It lost $147 million in the first nine months of 2000. By then, the bloom was off the rose, and Pets.com could not raise money to stay in business. The company folded in November 2000, laying off 300 employees.

Another spectacular failure was Webvan, an online grocery delivery business. In November 1999, Webvan raised $375 million in an initial public offering, giving a company with inception-to-date revenue of less than $400,000 a market valuation well over $1 billion. When investors started looking closer, they realized that the numbers were not going to work. When Webvan failed in July 2001, it laid off 2,000 employees.

Founded in 1997, eToys.com became the "go-to" holiday shopping site of its day. Its stock hit a high of $84.35 per share in October 1999. However, with millions spent on marketing and tough competition from Toys"R"Us, Amazon, and Walmart, eToys.com couldn't keep up. They lost $74.5 million in the fourth quarter of 2000—the all-important holiday months for retailers—and filed for bankruptcy in February 2001.

BUBBLE WITHIN A BUBBLE—TELECOM

To bring out a new technology for consumers first, you just had a very long road to go down to try to find people who actually would pay money for something.
—MARC ANDREESSEN

The growth of data traffic fueled by the rise of the Internet also led to a boom in investment by telecommunication companies like Global Crossing and WorldCom. They invested hundreds of billions (mainly financed with debt) into wireless networks, switches, and fiber optics. All of this new capacity eventually outstripped demand. The cost of bandwidth collapsed despite internet usage doubling every few years. When the telecom bubble burst, these companies (and others) imploded under the weight of all that debt.

Global Crossing was founded in 1997 and soon went on an investment and acquisition binge. At its peak, it was valued at $47 billion during the dot-com bubble. Global Crossing invested $15 billion in building fiber-optic networks around the world. Throughout its existence, the company never made a profit. In the fourth quarter of 2001, it had $793 million in revenue yet still managed to lose $3.4 billion. It filed for bankruptcy in January 2002.

WorldCom was another large telecommunications company that went bankrupt in 2002. Its market capitalization peaked at $175 billion. When the bubble burst, companies slashed spending on telecom services and equipment, and WorldCom resorted to accounting tricks and fraud to maintain the appearance of ever-growing profitability. Investors began to suspect that this might be too good to be true, and by July 2002, WorldCom became one of the largest bankruptcies in U.S. history.

In January 2000, at the frothiest phase of the bubble, America Online (AOL) merged with Time Warner. This was the largest merger in American business history at the time, valued at $350 billion. Combining Time Warner's huge book, magazine, television, and movie production assets with AOL's 30 million Internet subscribers seemed like a brilliant idea, but the timing could not have been worse.

Cheap broadband was starting to change the media industry's landscape. AOL dialup users migrated to high-speed internet access. As the dial-up user base shrank, AOL Time Warner never found the synergy it sought. One year after the merger, the combined entity lost $99 billion. The stock price for both companies went down 90% by the end of 2002. This was considered by many to be the worst merger of all time.

Not every dot-com failed, however. Companies that survived and prospered after the bubble burst had the right people in place, a solid business plan, a defensible niche in the marketplace, and just the right amount of good luck. Amazon.com, for example, sold $672 million in convertible bonds to overseas investors in February 2000. Amazon lost 90% of its value in the crash and would have almost certainly faced insolvency without that well-timed financial cushion.

BUBBLE WITHIN A BUBBLE WITHIN A BUBBLE—BEANIE BABIES!

> *You can make a lot of money with a good cat.*
> —TY WARNER

One notable dot-com success story is the online marketplace eBay.com. Its early success was mainly because it became the place to go for the latest hot toy, Beanie Babies. In May 1997, eBay sold an estimated $500 million worth of Beanie Babies! The company went public in September 1998 with a target price of $18 per share. The stock closed at $53.50 on the first day of trading. By the end of that year, Beanie Babies accounted for 10 percent of eBay's revenue.

Ty Warner launched a line of colorful, inexpensive plush toys called Beanie Babies in Chicago in 1993. As sales began to take off, Warner cleverly created a market buzz and a sense of scarcity. He would retire certain Beanies, upping the ante on the secondary market (e.g., eBay). Items that initially sold for $5 soared into the hundreds and even thousands of dollars. The Princess Diana bear became the Holy Grail of Beanie Babies, with offers on eBay reaching five figures.

An entire market ecosystem emerged for Beanie Babies, from eBay to magazines to trade shows. The toys were mass-produced, however, so by the turn of the century, only a few were rare. Prices eventually collapsed. The Beanie bonanza is eerily reminiscent of the Tulip Bulb Mania nearly four centuries ago. Today, neither tulips nor Beanie Babies are the exclusive domain of the rich.

18

U.S. HOUSING BUBBLE AND GLOBAL FINANCIAL CRISIS (2007-08)

A snarky but accurate description of monetary policy over the past five years is that the Federal Reserve successfully replaced the technology bubble with a housing bubble.
—PAUL KRUGMAN

THE HOUSING BUBBLE had roots going back decades. With broad support from the American electorate, politicians from both sides of the aisle have long promoted legislation to support the American dream of home ownership.

At Congress's direction, the U.S. Department of Housing and Urban Development (HUD) required lenders to increase the share of their loans to low- and moderate-income households. To accomplish this, lenders reduced down payment requirements, lowered credit standards, and extended more loans to subprime borrowers (who were more significant credit risks). Subprime mortgages rose from 4.5 percent of loans originated in 1994 to 20 percent in 2006.

An extended run of low interest rates and rising house prices from 2002 to 2005 prompted many individual investors to try to make a living buying and selling houses on borrowed money. Mortgage originators aggressively promoted NINJA (No Income, No Job Application) loans. The 2015 movie "The Big Short" told the story of a stripper who owned five houses and a condo obtained almost entirely on borrowed money, convinced that the property's value would only go up and her adjustable-rate mortgages could be easily refinanced.

Mortgage originators quickly turned around and sold their shaky loans, locking in their profits and shielding themselves from future defaults. Large financial institutions like Lehman Brothers, Bear Stearns, and Goldman Sachs bundled the loans and sold them as "safe" mortgage-backed securities to unsuspecting clients.

Many of these loans were Adjustable-Rate Mortgages (ARMs) with introductory rates that expired after two or three years. Adjustable-rate mortgages jumped from 11 percent of total outstanding mortgages in 2002 to 22 percent in 2008.

Lenders also targeted existing homeowners for new loans. Borrowers were encouraged to refinance to take advantage of lower interest rates and to extract home equity. Brokers lowered their underwriting standards while peddling ever-riskier products. Many homeowners began to view their home equity line of credit as an unlimited cash machine available for household expenses and discretionary spending such as automobiles or vacations.

THE BUBBLE BURSTS

It has often been said that the Federal Reserve's job is to know when to remove the punch bowl at the party, so it dutifully began raising short-term interest rates in 2005. Housing prices declined, and adjustable-rate loans began to reset on a massive scale. Defaults and foreclosures quickly followed.

By 2011, Arizona median home prices had dropped more than 50% from their 2006 peak. MarketWatch estimated in 2011 that nearly two-thirds of Phoenix area mortgages were underwater (meaning that the amount owed was more than the house's total value). Owners with no equity (or negative equity) walked away, leaving banks and investors holding the bag.

Banks and investment firms holding mortgage-backed securities began to realize that they had a severe problem. Nearly $5 trillion of mortgage loans were in trouble, and their capital reserves were woefully inadequate to absorb the losses. Mortgage-backed securities collapsed in value, and financial institutions worldwide suffered severe damage.

Lehman Brothers was the first to go. In September 2008, they became the largest bankruptcy filing in U.S. history, involving over $600 billion in assets. Regulators quickly realized that other big banks had similar problems and would suffer the same fate unless the government intervened.

In September 2008, the Federal Housing Finance Agency (FHFA) took over loan guarantors Fannie Mae and Freddie Mac. The Federal Reserve guaranteed loans for Bear Stearns and took over American International Group (AIG) with $85 billion in debt and equity funding. Washington Mutual was seized by the Federal Deposit Insurance Corporation (FDIC) and went bankrupt when there was a run on the bank.

U.S. Secretary of the Treasury Henry Paulson and Chair of the Federal Reserve Ben Bernanke requested a $700 billion fund to acquire toxic mortgages. They warned: "If we don't do this, we may not have an economy on Monday." The U.S. Congress passed, and President Bush signed the 2008 Emergency Economic Stabilization Act.

Altogether, central banks worldwide purchased $2.5 trillion of troubled assets from banks in the fourth quarter of 2008. This was the largest liquidity injection into the credit market in world history. Even with this huge capital injection, the FDIC closed 450 banks between 2008 and 2012, compared with a more typical failure rate of just a couple of banks per year. Cadres of men and women in dark suits would descend on banks on a Friday afternoon, locking out employees and reopening on Monday under new ownership. Yes—they are from the government, and No—they are not here to help you.

When your friends quit their day jobs to flip houses and exotic dancers own more investment properties than you do, it should be a sign that delusion has set in and you are in a bubble destined to burst.

19

GHOST CITIES—THE CHINESE HOUSING BUBBLE

What makes China so opaque—and, indeed, what gives the government such control over information in the first place—is that its rules are fluid.
—DINNY MCMAHON

THE PEOPLE'S REPUBLIC OF CHINA IS the world's most populous country, with more than 1.4 billion people. Over the last three decades, China has experienced blisteringly high economic growth rates, leading to an ever-growing middle class hungry for better lives and homes.

Real estate is one of the few investment vehicles available to ordinary Chinese citizens, accounting for seventy percent of personal wealth. If you are a young man and own a home, you are considered good marriage material. If you have children, you need a home to get them into good schools. Chinese home buyers are willing to pay thirty to fifty percent upfront for unfinished projects, viewing this as an investment and a future place to live for themselves or their children.

The housing bubble in China developed when local governments started using land sales to generate revenue. They rezoned low-value collectively-owned farmland into high-value multi-family housing land. Then, they sold this land to developers who would build large-scale housing projects, sometimes irrespective of demand.

The problem was compounded by a slowdown in economic growth in 2019 and 2020 and worsened by the COVID-19 pandemic. Demand for residential housing collapsed. Developers counted on the continued infusion of new cash and loans to complete their projects. When home buyers stopped buying and banks stopped lending, many projects languished unfinished. Developers had millions of empty homes and unfinished units where no one wanted to live.

The largest developer to fail was Evergrande. In 2020, it had land agreements and financial leverage to build housing for 10 million people. It felt flush enough to expand into other markets, such as electric cars, theme parks, and even a soccer club. By the summer of 2021, rating agencies began to warn of liquidity problems. By the fall of that year, Evergrande started defaulting on bond payments, forcing the Chinese government to intervene. Trading of Evergrande's shares was halted in January 2022.

Evergrande was not alone. Its liquidity crisis and collapse also resulted in closer scrutiny of other property developers. This led to even more construction stoppages and angry home buyers across China.

In some areas (mainly large cities), demand for housing still outstrips supply. But there is a glut of empty developments in rural areas. Block after block of high-rise buildings in the wrong places sit empty and unfinished, and shopping centers remain empty. These have come to be called ghost cities and phantom malls.

One prominent example is the Kangbashi District, 25 kilometers south of the city of Ordos (population two million) in the desert region of Inner Mongolia. Kangbashi was initially intended to accommodate another million people by 2023. Developers built a downtown area with government buildings, a museum, an opera house, a library, and over 40,000 apartments. Eerie photos make the Kangbashi District look like a glittering new city, except that it remained nearly empty for years. Due to the cost of housing and lack of urban infrastructure and services, people commuted from nearby Ordos rather than moving to Kangbashi.

Over 150,000 people live in the Kangbashi District today, so it is not quite a ghost town, but eighty to ninety percent of the presold apartments still sit empty. People purchased them as retirement homes, homes for their children when they married, or long-term investments. Many appear from the outside to be completed, but the interiors are empty concrete shells.

In 2011, there were an estimated 89 million empty homes in China. In 2020, Business Insider estimated that number remained as high as 65 million. However, markets in China are somewhat distorted by government control, and the government tends to take the long view. It may be decades before the dystopian landscape of empty high-rises, expansive parks, and grand monuments is populated and begins to flourish.

20

CRYPTOCURRENCY — THE REEXAMINATION OF MONEY

A cryptocurrency is not a currency, not a commodity, and not a security. Instead, it's a gambling contract with a nearly 100% edge for the house.
—WARREN BUFFETT

CRYPTOCURRENCY (also known as crypto) is simply an electronic medium of exchange. Instead of paper and coins issued by government decree, cryptocurrency is a digital key (a string of letters and numbers) along with some associated unit of value (for example, Bitcoins or dollars) issued and traded over the internet using cryptography for privacy and security.

Is cryptocurrency delusional? Money is whatever we all agree it is. It could be beads, precious metals, salt, paper, or electrons. As long as we remain confident that we can trade it for goods and services, it works for us as money. Like any other currency, the value of cryptocurrency depends entirely on agreement between willing buyers and sellers as to the value of those coins. In other words, cryptocurrency is not delusional as long as someone is willing to pay good money for it.

For much of human history, money was a commodity, where the medium of exchange had intrinsic value (such as salt or precious metals). Later, representative money became more dominant. With representative money, the unit of exchange has no inherent value, but it represents and is supported by something that does. U.S. Silver Certificates were an excellent example: each certificate was backed up by a fixed amount of silver bullion sitting in a government vault. Today, most global currencies are classified as fiat money, issued by a government and not backed by any commodity but instead by the faith and credit of the issuing body and confidence that others will accept the currency.

But some types of money are better than others. Good money is widely accepted, easy to trade, and retains its value over time. Bad money fails on one or more of these measures. When cryptocurrency is put to the good money / bad money test, it fails. Cryptocurrency is not widely accepted, is easy to trade only if you are technologically adept, and its ability to retain value over time is highly suspect. If that sounds like bad money to you, then you understand cryptocurrency perfectly.

BITCOIN

The most widely accepted and recognizable cryptocurrency of our time is Bitcoin. Bitcoin was launched in 2009 by an unidentified person using the pseudonym Satoshi Nakamoto. Nakamoto envisioned Bitcoin as a global alternative to traditional fiat money. His goal was for Bitcoin to eventually become a globally accepted legal tender and be widely used to purchase goods and services. Nakamoto created software to support this new idea of electronic money that contains a distributed public database for recording transactions, tools for storing and trading coins directly between users, and a mechanism for issuing new coins and validating transactions while limiting the total supply of coins (Bitcoin mining).

Nakamoto started with one Bitcoin in 2009 and designed a "Bitcoin mining" process to create new coins in progressively smaller amounts over time. He artificially capped the total supply at 21 million coins. Experts predict that the last Bitcoin will be mined by 2040.

To use Bitcoin to pay for a product or service, users supply a personal digital key from their "Bitcoin wallet" to complete the transfer. Bitcoin wallets are software that provides the functionality to secure, send, and receive the cryptographic keys that prove ownership of a specific amount of value on the Bitcoin network. When transactions are added to the database, a public record of the exchange remains there indefinitely.

The Bitcoin database uses a technology known as "blockchain" to allow transactions to be verified, grouped in "blocks," and added to the "chain" in a public database. As new transactions occur, new blocks of transactions are permanently added to the database.

Bitcoin miners are rewarded with a new Bitcoin for solving a cryptographic puzzle that, when completed, lumps together the next block of transactions and adds them to the public database. Solving this puzzle requires an enormous amount of specialized computational power. The degree of difficulty of the cryptographic puzzle is adjusted over time based on the number of Bitcoin miners and the computational power being applied. This keeps the rate of new coin generation at roughly one coin every ten minutes.

Bitcoin mining can be lucrative. Entire data centers have been built to mine Bitcoins. According to the Cambridge Bitcoin Electricity Consumption Index, global electricity usage for crypto mining exceeds 120 billion kilowatt-hours annually, more than that of countries like Austria, Chile, and Israel!

BENEFITS OF BITCOIN

If you can make the intellectual leap that digital keys have value because the mysterious programmer who created them tells you they are scarce, some aspects of this new currency can be very appealing. First, the risk of currency devaluation from the government printing more money is

CRYPTOCURRENCY—THE REEXAMINATION OF MONEY

eliminated. The money supply is known and fixed.

Another appealing characteristic is that Bitcoin can easily be traded directly between users. The ability to trade without involving a third party provides anonymity unavailable from traditional banks or financial institutions. Plus, you can hold, send, or receive tiny fractions of a Bitcoin. This easy divisibility has become a key selling point as the value of Bitcoin has soared into the tens of thousands of dollars per coin.

Bitcoin transactions can be trusted because every transaction is logged in a publicly accessible, secure, and decentralized database. All users collectively have equal access to the data. Blocks are only added and never deleted, so no single person or group (or government) can alter or control the information in the database.

Finally, cryptocurrency may be a good alternative for the unbanked, especially in developing nations. If there are no reliable financial institutions where you live, but you have a cell phone with internet access, you can participate in global markets on a level playing field using cryptocurrency. In economies beset by hyperinflation, cryptocurrency may serve as an excellent store of value compared to rapidly depreciating local currency.

So, what could possibly go wrong with a brand-new global currency created by a shadowy figure hiding behind a pseudonym? Quite a lot, as it turns out.

ANONYMITY AND CRIME

One of the main reasons for creating Bitcoin was to allow users to trade directly with each other and eliminate the need for a bank or other third party to execute the transaction. This feature is uniquely well-suited for money laundering, extortion, drug dealing, and other criminal purposes. Bitcoin buyers and sellers don't use their names in transactions; they create a code that is their digital signature in the blockchain. The transaction itself is public, while the identity of the Bitcoin owner remains hidden.

Fortunately, law enforcement has figured out that they can use the transaction database to isolate the user's internet address. This information can often be traced back to specific computers and individuals. The paradox of cryptocurrency is that while your transactions are "peer to peer" (directly between users) and securely encrypted, the transaction database creates a forensic trail that can make your entire financial history public once you are identified as the account owner.

The best example of this was the Silk Road online marketplace. In 2011, Silk Road sold and distributed massive quantities of illicit goods and services. The FBI traced Silk Road founder Ross Ulbricht to a coffee shop in California. Logs on his computer identified him as the Silk Road system administrator. He was arrested by the FBI in 2014 and sentenced to life in prison in 2015.

Ironically, Silk Road was also the victim of the theft of 50,000 Bitcoins in 2012. James Zhong created multiple Silk Road accounts and then triggered dozens of transactions in rapid succession. The built-in delays in Bitcoin mining and posting transactions tricked Silk Road's withdrawal-processing system into releasing over 50,000 Bitcoin into Zhong's accounts. These funds were quickly re-distributed to other accounts. The U.S. Attorney and the IRS announced a historic $3.36 billion seizure in November 2022. James Zhong pled guilty to committing wire fraud in this theft.

Bitcoin is also the currency of choice for ransomware. Ransomware is malicious software that encrypts files on a computer and makes them unusable. The hacker then demands a ransom payment in Bitcoin to decrypt the victim's files. Ransomware attacks are estimated to occur every 11 seconds and generate over $1 billion in revenue annually for cybercriminals.

CRYPTOCURRENCY EXCHANGES

The need for liquidity in the cryptocurrency marketplace has led to the creation of hundreds of exchanges. Like a stock or bond market, exchanges enable you to buy or sell cryptocurrency for dollars or Euros. Exchanges vary widely in their features, fee structures, and the level of risk they introduce.

FTX Trading Ltd., founded in 2019, was, at one point, the third-largest crypto exchange in the world. At its peak, it had over a million users and commanded a market valuation of $32 billion. FTX sponsored the Miami Heat's basketball stadium, renaming it FTX Arena, and partnered with Major League Baseball to place the FTX logo on umpire uniforms. It donated millions to politicians and obtained celebrity endorsements from NFL quarterback Tom Brady, MLB slugger David Ortiz, and NBA star Steph Curry.

FTX collapsed in November 2022 when it experienced a run on the bank. Founder and CEO Sam Bankman-Fried had been using money from FTX investor accounts to pay for other hedge fund investments. When word of this got out, everyone wanted to cash in their accounts simultaneously, and liquid assets were not there to fulfill those requests. This triggered a collapse in the value of a single Bitcoin from over $60,000 to less than $17,000.

Amazingly, FTX's board of directors and senior executives consisted mainly of the founder's college friends. This $32 billion firm used the small business accounting software Quickbooks. Seasoned venture capital funds invested nearly $2 billion into the company without a single board seat in return. Sequoia Capital, for example, wrote down its equity in FTX after the collapse from $214 million to $0.

John J. Ray III, the CEO brought in by the court to manage the business during bankruptcy, testified before a U.S. Congressional committee that despite having been involved in other high-profile bankruptcies (including Enron, Residential Capital, and Nortel), "Never in my career have I seen such a complete failure of corporate controls and such a complete absence of trustworthy financial information." Ultimately, FTX's founder, Sam Bankman-Fried, was convicted of fraud and sentenced to 25 years in prison.

DON'T LOSE YOUR WALLET

Another cryptocurrency risk is that coin holders can lose access to their accounts. If a user with an independent wallet (i.e., not on an exchange) were to have their wallet compromised, forget their password, or die without sharing their wallet details, any value associated with that account would become irretrievable. No central bank or financial institution is available to reset your password.

San Francisco software developer Stefan Thomas was an early adopter of Bitcoin who managed to amass over 7,000 Bitcoins. He stored his Bitcoin on a "hardware wallet" that limits users to ten wrong password guesses before it encrypts the contents permanently. And then he forgot his password. At last report, he had just two guesses left to access Bitcoin worth over $100 million. If he gets it wrong, that money is gone forever. And he is not alone. According to a 2020 study by cryptocurrency data firm Chainalysis, as much as 20 percent of the Bitcoin already issued (tens of billions of dollars) may be permanently lost.

VOLATILITY

Like all types of money, the price of Bitcoin depends on supply and demand. The mining process throttles the Bitcoin creation rate, and there is a cap of 21 million coins. So, the supply side is quite transparent. The demand side is another thing altogether. Any hype, excitement, or fear created by governments, influencers, or industry moguls can lead to significant swings in the price of Bitcoin.

Michael Saylor, the CEO of MicroStrategy (aka The Bitcoin Whale), has caused price spikes by promoting the currency on social media. Cryptocurrency advocate and Tesla CEO Elon Musk frequently tweets about digital currencies, resulting in soaring and falling price movements. At one point, Tesla began accepting Bitcoin from customers to buy cars. This offer was later withdrawn due to environmental concerns about heavy energy usage associated with Bitcoin mining.

In September of 2021, El Salvador became the first country to adopt Bitcoin as a legal tender, requiring businesses to accept it as payment. President Naybi Bukele announced the news on Twitter, saying, "In 3 minutes, we make history." The price of Bitcoin increased by 1.49 percent on that day. It fell by more than half in the months that followed.

One day, the last Bitcoin will be mined. Current projections put that date at 2040. What happens, then? Bitcoin miners have been rewarded for their efforts with Bitcoin itself and the transaction fees for the block they mined. When there are no more coins to be mined, transaction fees will have to increase substantially to provide enough incentive for adding blocks to the blockchain. If not, transactions will take longer to update or not happen at all.

COMPETITION

Bitcoin was the original cryptocurrency, but thousands of others have popped up. Some are mere copycats, and others attempt to carve out a niche or improve on areas where Bitcoin is weak. Some set themselves up as stablecoins, backed by a fiat currency like the Euro or the U.S. dollar. Others use different technologies to validate transactions more quickly than Bitcoin. Dogecoin (dodgy coin?) started as a joke in 2013 but later evolved into a real cryptocurrency. Unlike other cryptocurrencies, there is no cap on the number of Dogecoins that can be created, which leaves it highly susceptible to devaluation.

The idea of directly connecting savers and borrowers on a low-cost digital platform has spurred central banks to consider creating digital versions of their currencies. China, Japan, and Sweden are already conducting trials in this arena, and multinational firms like Meta (Facebook) and JP Morgan are also testing the cryptocurrency waters with their own versions.

Just 15 years after the introduction of Bitcoin in 2009, there are more than 12,000 different cryptocurrencies, an astonishing growth rate. There are millions of crypto users worldwide and hundreds of crypto exchanges. While these currencies don't appear to threaten the Euro or the U.S. dollar, they could potentially undermine the currencies of smaller and less developed countries.

Some believe that a scarce digital asset like Bitcoin could one day replace the U.S. dollar as the global reserve currency. But scarcity alone may not be enough. Price volatility, market instability, and competition from new entrants make Bitcoin look like a speculative bubble. Investors who come to the party late are at greatest risk of losses. Trillions of dollars of electronic money depend entirely on our willingness to accept crypto as payment for goods and services. It works when we all agree. It becomes bad money and a delusion when we don't.

21

THE BERNIE MADOFF PONZI SCHEME

It's a proprietary strategy. I can't go into it in great detail.
—BERNIE MADOFF

IN A PONZI SCHEME, "too good to be true" investment returns are paid using money raised from new investors. The belief that values can only go up is similar to a bubble, but the entire basis of the scheme is fraud. To the moon, Alice!

Ponzi schemes are a crime, but do they qualify as delusion? Like many other types of fraud, financier Bernie Madoff's Ponzi scheme started small. Early investors in his firm were family friends, so he felt enormous pressure to achieve the promised returns and believed he could make up any losses with future trading. Even years into the fraud, he still believed he could recoup the losses for his investors. This was delusional thinking, as the hole he dug only got deeper.

Madoff founded a penny stock brokerage right out of college in the early 1960s. His well-connected father loaned him money and referred several customers to help him get started. Bernard L. Madoff Investment Securities, LLC became a top "Market Maker" (buyer and seller) of thinly traded Nasdaq and S&P 500 stocks. The firm also had a less well-known investment management division that was the heart of the Ponzi scheme.

The mechanics of the fraud were simple. When a customer chose to invest, Madoff determined the returns each customer should earn, and then created reports with false trade dates and amounts to reflect the gains he had promised. Occasionally, trades shown on the false statements were on dates when markets were closed! Unfortunately, this was not noticed until it was too late.

The money went into a pooled JPMorgan Chase investment account. When someone needed to make a withdrawal, it came out of the pool of other customers' money in the account.

With every account performing as promised, referrals and testimonials led to new investments, outpacing eventual customer withdrawals. Madoff was well-connected in wealthy American Jewish communities and used those connections to solicit investments from Jewish individuals and institutions. Jewish charities such as the Los Angeles Jewish Community Foundation, Foundation for Humanity (Elie Wiesel's charity), and the America-Israel Cultural Foundation lost millions of dollars.

The reported value of funds invested with Madoff when the scheme was exposed was $65 billion, although actual losses of invested funds were closer to $18 billion. This was the largest Ponzi scheme in American history. The real tragedy of this story was the regulatory failures that could have exposed the fraud years earlier.

Between 1992 and 2008, the U.S. Securities and Exchange Commission (SEC) investigated Madoff six times. None of these investigations uncovered the fraud or resulted in any regulatory action.

Even when presented with evidence, the SEC ignored whistleblowers and botched its investigations. Madoff later said he should have been caught in 2003, but the investigators never even looked at his stock trading records. The fraud would have been immediately obvious had they taken this basic step.

In 1999, securities industry executive and financial fraud investigator Harry Markopolos uncovered evidence suggesting that Bernie Madoff's wealth management business was a Ponzi scheme. He informed the SEC that achieving the gains Madoff claimed to deliver was statistically impossible. Securities investments do not go up 100% of the time. Even highly successful funds go down occasionally before finding their way back up. The SEC ignored Markopolos or gave his evidence only a cursory investigation despite multiple attempts to press the case. After Madoffs' conviction, Markopolos published a book called "No One Would Listen: A True Financial Thriller." Thriller indeed.

In December 2008, Madoff faced $7 billion in redemption requests and had only $234 million in his JPMorgan Chase account. He knew there was no way to borrow or raise the money he needed and that the game was over. He planned to distribute $18 million in early bonuses before word of the fraud leaked out, not realizing that regulators would immediately claw back those distributions. He confided in his sons, who quickly recognized that their choice was to either report their father for the crime or become accomplices after the fact. They chose to report the fraud to federal authorities.

Madoff pleaded guilty to 11 federal felonies without a plea bargain in March 2009 and was sentenced to 150 years in prison. He insisted that he was solely responsible for the fraud, although others were also later convicted in the scheme. Madoff died at a Federal prison in North Carolina in 2021. The U.S. government eventually collected and paid out $772.5 million to more than 24,000 victims of his Ponzi scheme.

Madoff's initial delusion was a small one. He believed he could recoup the gains on fudged performance reports and extricate himself from the scheme. Over time, his level of self-deception grew until he crossed the line to full-fledged criminal behavior. Investors indulged in delusional thinking as well. Madoff was a charismatic salesman, so they set aside doubts about his unbelievable track record of success because they wanted to believe.

22

WHOLE LOTTO DELUSION

There are many harsh lessons to be learned from the gambling experience, but the harshest one of all is the difference between having Fun and being Smart.
—HUNTER S. THOMPSON

THE URGE TO GAMBLE is as old as humanity. Some scholars believe that lotteries helped finance the construction of the Great Wall of China!

And while gambling can lead to addiction, it can also serve as a harmless form of entertainment. In either case, there is no doubt that there has been extraordinary growth in legal gambling throughout the United States over the last few decades. This growth has come from online gaming, the development of Native American casinos, and the decision by many states to use gaming (especially pull tabs, lotteries, and sports betting) as a source of revenue.

There is a continuum of delusional thinking associated with gambling. A bet on your favorite sports team, an occasional lottery ticket, or an evening at a casino can be simple entertainment. Gambling addiction lurks at the other extreme, where compulsion and magical thinking can overwhelm rational behavior.

The vast middle ground of gambling delusions comes from selective memory (remembering the wins, not the losses), superstitious beliefs, and misreading the odds. The worse the odds, the greater the delusion. State-run lotteries especially take advantage of this tax on being bad at math.

Lotteries are random number drawings for a prize. The range of numbers, the number of draws, and other factors can dramatically affect the odds of winning. A six-number draw from 49 balls offers a one in 14 million chance of success. A popular multi-state lottery in the United States selects five random balls numbered from 1 to 70, plus one bonus ball (numbered 1 to 25), resulting in 302 million to one odds of getting all six numbers correct. Six numbers are selected from ninety balls in the Italian national lottery, driving the odds up to a one in 622 million chance for success.

The purchase of a lottery ticket enables the buyer to indulge in dreams of becoming wealthy. In effect, people buy permission from themselves to indulge in a fantasy they would not have otherwise. But the odds of winning are so tiny that Fran Lebowitz once said, "I've done the calculation, and your chances of winning the lottery are identical whether you play or not."

By comparison, the odds of getting a perfect 29-point hand in cribbage are one in 216,580. The odds of being attacked by a shark are one in 4.3 million. The lifetime odds of being struck by lightning are one in 15,300, and the odds of being struck by lightning twice are one in 9 million.

Perversely, the more lottery tickets you buy, the worse your odds. Approximately sixty percent of lottery revenue is returned as prize money to the winners. The balance goes to proceeds for the sponsoring organization and covers costs for hosting the drawing. If you purchased all the tickets for a given drawing, you would immediately lose forty percent of your investment, even though you won all the prizes. The more a lottery ticket buyer risks indulging in this fantasy, the greater the delusion.

Lottery-related scams only compound the problem. Scammers solicit fees to claim winnings that don't exist, or they want you to pay for advice or sell fraudulent systems for picking winning numbers. Some scammers will say you are a winner in a foreign lottery that you did not enter. People who rely on fortune cookies or a pseudoscience like astrology to select lucky numbers for the lottery are doubly steeped in delusion.

23

MULTI-LEVEL MARKETING

Probably the worst thing about a multi-level marketing pyramid scheme is how it makes you advertise to all your friends and family that you fell for it.
—ANONYMOUS

MULTI-LEVEL MARKETING (MLM) is a business model where compensation is paid to participants from direct sales of products or services plus commissions generated by other sellers the participant has recruited. MLM can be a pyramid scheme, but it isn't necessarily. The Federal Trade Commission (FTC) distinguishes between legal multi-level marketing and illegal pyramid schemes based on upfront money required to participate and the level of emphasis on recruiting versus direct selling.

Credit for inventing the concept of multi-level marketing goes to a company called Nutralite in the 1940s. Early in the company's history,

Nutralite distributors realized they could generate additional sales by recruiting new people to sell on their behalf. Once the company settled on a reimbursement model of paying the recruiting sponsor a two percent commission for downline sales, unprecedented growth and profits followed.

Two successful Nutralite distributors eventually left and started Amway. In 1956, another Nutralite distributor left to form Shaklee, specializing in nutritional supplements and cleaning products. Amway eventually absorbed Nutralite into its operations, and Shaklee became a Fortune 500 company in its own right. More than 600 multi-level marketing companies are active in the United States today.

Direct selling is much more difficult than most people realize, especially once you have made your pitch to family and friends and need to start cold-calling strangers. Recruiting multiple layers of "downline" salespeople who will help you become rich is even more difficult.

In 2021, Amway charged an annual registration fee of $76 for its distributors, who are called Independent Business Operators or IBOs. IBOs earn money on the markup from their product sales plus a bonus or commission on product sales made by their sales team (or downline). According to the 2021 Income Disclosure document on the Amway website, the top fifty percent of IBOs earned an average annual income of $3,414. The median annual income of this same leading group was $631, meaning that half of Amway representatives earn less than $53 per month! Amway promotes being in business for yourself but not by yourself. $53 per month is not much of a business.

Perversely, Amway has had enough success over the years that almost everyone has heard of it. Amway's distributors and recruiters are trained not to identify the company by name until several meetings into the recruiting sales pitch. They worry that prematurely mentioning the company would elicit an immediate and hard "No."

A 1980 investigation of tax returns by the Wisconsin Attorney General showed that while two Amway Direct Distributors in the state earned over $50,000, the average net income after subtracting operating expenses for the top one percent of Amway distributors was minus $900. That's right, they lost money. A 2018 survey of 1,049 multi-level marketing participants conducted by MagnifyMoney found that most made less than 70 cents per hour before deducting expenses.

Another study of MLM earnings based on tax return data shows that Amway IBOs are not alone in their poverty. If you were a multi-level marketing business owner in Utah in 2002, you had a 97.9 percent chance of losing money. You could do better in Las Vegas.

One major factor contributing to financial losses is inventory loading. Since the FTC considers recruitment-only MLM an illegal pyramid scheme, companies typically require their representatives/recruiters to sell products. People focused on the recruitment aspect of their business buy products they may not need to remain eligible to earn commissions or bonuses from their downline, often winding up with a closet full of unneeded products.

Multi-level marketing companies sell the dream of financial independence, knowing that most of their recruits will fail. With this ethical lapse at their core, it is a short step to making false and misleading health and product benefit claims. In 2023, the FTC issued warning letters to almost 700 companies that they could face civil penalties if they couldn't back up product claims on their labels. Six hundred seventy of those companies market over-the-counter drugs, homeopathic products, dietary supplements, or functional foods using a multi-level marketing business model.

These companies have learned not to make unsupported health and medical claims on their websites or marketing materials. However, their independent sales representatives may be unaware of or unconcerned about federal regulatory requirements. They quickly figure out that the idea of disease being cured and health restored sells, so they make their claims using informal settings such as home parties, networking groups, and other face-to-face methods, which are much more difficult for regulators to detect.

The delusion of multi-level marketing is believing the hype that everyone involved is a winner. Unfortunately, the entire system is built on a foundation of deception. The only winners are the company's owners and the tiny percentage of distributors at the top of the sales pyramid.

24

TIMESHARES

It's morally wrong to allow a sucker to keep his money.
—W. C. FIELDS

TIMESHARES ARE VACATION PLANS that consist of partial ownership of a vacation property. You pay an upfront price (plus an annual maintenance fee) to purchase a share of a unit in a vacation property. This gives you access to the property for a period of time (generally sold in blocks of one week), usually for the same time slot each year. According to the American Resort Development Association (ARDA), this is an $8 billion industry with more than 1,500 timeshare resorts in the U.S., averaging 132 units per resort.

The delusion of timeshares is that they are sold in high-pressure sales environments as an asset or an investment that will appreciate and save you money in the long run. They are none of these things.

According to the ARDA, the average cost for a one-week timeshare today is over $24,000, plus annual maintenance fees ranging from $640 to $1,290. Most timeshare agreements obligate you to pay the yearly maintenance fee indefinitely, even if you cannot use your week. Additional fees apply if you "bank" your week (save it for use at another time) or trade for a different date or time at another resort.

Let's take a look at the math. It might cost a developer $40 million to buy land and build a high-end 130-unit vacation resort in the United States. That developer would have 52 timeshare weeks to sell for each unit or 6,760 total timeshare units. At $24,000 per unit, the property would generate over $162 million if every share were sold. If the developer only ever sold half of the available share, they would still double their money. A $1,000 annual maintenance fee would cover operating costs of $6.76 million per year, going up over time as needed.

Let's also look at the math from the buyer's perspective. Your $24,000 ownership share buys you one week per year at that resort forever. If you were to apportion the initial cost over forty years, your annual cost would be $600 per week. But you must also pay $1,000 for the yearly maintenance fee, so your expenses are $1,600 ($600 plus $1,000) for each vacation week.

If you rent a comparable hotel room in the same area, you might pay $200 per night ($1,400 for the week). If your timeshare costs for a one-week vacation are $1,600, and a nearby hotel is only $1,400, you are overpaying $200 per vacation week to start. Your losses increase over time as the annual timeshare maintenance fees increase due to inflation.

All of that assumes that you want to and can go to the same place the same week of the year for forty years. Tough luck if you miss a week. If you cancel or change your week, there are fees. If you trade for a different timeshare location, there are fees. If you "bank" a week, there are fees. These fees can easily cost hundreds of dollars per change. Many timeshare operators have switched to a "points" or "credits" model to deliberately obfuscate the value of your share and the amount of your losses. Effectively, they are making your timeshare even more abstract to own.

The actual value of timeshare properties can be more accurately gauged by looking at the resale market. People trying to sell their timeshares quickly learn that there are a lot more people wanting to sell than wanting to buy. The resale market is so huge that the likelihood of recovering anywhere near your initial investment is practically zero.

Hundreds of timeshare resale websites populate the internet. Some fraudulently collect up-front fees and then fail to perform (adding insult to injury). Timeshares listed on Craigslist can range from one dollar to a few thousand dollars. Another popular reseller estimates that 85% of timeshare owners want to sell and tells sellers, "Even if you don't get any money at all, solely getting out from under those recurring fees is the best happily-ever-after you can expect."

High-pressure sales agents pitch timeshares as an investment that will grow in value. They want you to believe that your ownership share is an asset that can be handed down to future generations. However, the timeshare does not generate income. You must pay an ever-growing annual maintenance fee, so while it may bring you some enjoyment, it is more appropriately considered a liability.

Many timeshare sellers receive no offers and are left to pay hundreds or even thousands of dollars to have their name removed from the title and eliminate the annual maintenance fee obligation. Any asset class that adopts the "Pay when you buy and pay again when you sell" investment model can only be described as a delusion.

VOLUME 3

YOUR SOUL (CULTURAL, RELIGIOUS, AND POLITICAL DELUSIONS)

My land is bare of chattering folk; The clouds are low along the ridges, And sweet's the air with curly smoke; From all my burning bridges.
—DOROTHY PARKER

CHARLES MACKAY wrote about the delusional thinking that led to centuries of holy crusades to Jerusalem and the Middle East. He delved into witch burning, alchemy, astrology, magnetizers, and fortune tellers of the 18th and 19th centuries. We are no less delusional today with conspiracy theorists, faith healers, and suicide bombers.

Charles Mackay elected not to explore religious manias or delusions. He stated in his preface that a mere list of them would be sufficient to occupy a volume. I suspect he also calculated that he would deeply offend many of his readers. I shall savor the smoke of burning bridges and examine some of the more delusional political and religious thinking on display in modern society.

So, what is faith? Faith is simply accepting and believing in something you cannot empirically prove. Atheists deny the existence of God, yet they have no more basis for this belief than the faith they deny. They believe they are correct but can't prove it. They have only faith.

And what is delusion? Delusion is holding false beliefs despite overwhelming evidence to the contrary. Religious faith can't be delusional by definition because there is no incontrovertible evidence to the contrary for any religion. The strongest evidence against any particular set of religious beliefs may be the simple fact that many other people have different ideas.

People with religious faith will search for evidence that supports their beliefs. They fill the gap between evidence and dogma with a conscious decision to believe that which cannot be proven. They choose to have religious faith. People become deeply attached to their faith, some even to the point of martyrdom or murder.

Consider, however, that if one set of beliefs conflicts with another set of beliefs, one of them is at least partly wrong, no matter how deeply held those beliefs are. If you lay a hundred sets of conflicting beliefs side by side, only one of them (at best) can be entirely correct. The other ninety-nine will be wrong, at least to some degree. Some of them will be only slightly wrong, while others are entirely, outrageously, and obviously (to you and me) one hundred percent wrong.

In other words, mine is the one true faith, and the rest of the world is nuts. This volume will broadly outline several belief systems. Pick your favorite, and we can happily agree that the rest are delusional.

CULTURAL DELUSIONS

25

CONSPIRACY THEORIES — AN OVERVIEW

History is much more the product of chaos than of conspiracy.
—ZBIGNIEW BRZEZINSKI

CONSPIRACY THEORIES are an attempt to simplify a complex and confusing world. Bad things don't just happen; someone must be behind them. Conspiracy theories attempt to portray complex realities in broad and sweeping strokes with events driven by powerful, hidden, and evil forces.

Conspiracy theories tend to arise and catch on in times of societal crisis, such as political upheaval, terrorist attacks, plane crashes, natural disasters, or war. Feelings of fear, uncertainty, or being out of control stimulate a desire to make sense of one's social environment. People with a sense of powerlessness, unhappiness, or dissatisfaction with their

CONSPIRACY THEORIES—AN OVERVIEW

lot in life can be vulnerable to simple answers that help make sense of the world. They fall victim to attribution error, believing every incident must be intentional. Random events, bureaucratic incompetence, and unintended consequences are more difficult to accept than the idea of the Illuminati lurking in the shadows.

There is a great deal of overlap among conspiracy theory believers. It is not one group of people that believes the moon landing was a hoax and another that thinks mind control microchips are hidden in vaccines. It is, by and large, the same people.

The difference between a conspiracy theory and a scientific theory is that the scientific theory is filled with doubt, while the conspiracy theory is filled with certainty. Conspiracy theories are generally advanced with conjecture more than evidence and are often reinforced by circular reasoning. Every new bit of evidence against a conspiracy theory can easily be dismissed by claiming that even more people are becoming part of the conspiracy.

Conspiracy theories aren't new and are not exclusive to the United States. Consider the Illuminati. They were a secret society in Bavaria that operated for only a decade, from 1776 to 1785. Yet they are often viewed as the hidden hand behind many historical events, ranging from the French Revolution to the Battle of Waterloo and the assassination of John F. Kennedy. Many United States presidents and even Hollywood figures have been accused of belonging to the Illuminati.

Another example dates back to Roman times. In 64 AD, the great fire of Rome erupted while Emperor Nero was out of town. Conspiracy theorists of the time asserted that Nero and his associates deliberately started the fire and that he was singing and playing the fiddle while Rome was burning. Nero fiddles while Rome burns. Nero responded with his own conspiracy theory that the Christian community initiated the fire, leading many Christians to be persecuted and killed.

Conspiracy theories about plots to achieve world domination by large corporations and superpowers abound in Africa, Asia, and Eastern Europe. Russian society is saturated with conspiratorial thinking. Some

African conspiracy theories accuse conspirators of enacting their plans via sorcery or witchcraft. Chinese conspiracy theories about financial manipulation and climate change are often (weirdly) ascribed to the Rothschild family. Conspiracy theories about Jewish plots and the war against Islam fuel anti-Semitism throughout much of the Muslim world.

Conspiracy theories have mushroomed as a cultural phenomenon in recent decades with the rise of social media and the internet. Many studies have assessed the level of acceptance of conspiratorial beliefs. Survey and research results vary, but they cluster around a quarter to a third of the population for any given conspiracy theory. This is an astounding level of delusion, especially given the wide availability of reliable contradictory information.

In times of crisis, many conspiracy theorists start with the simple question: Who benefits? They deduce that whoever benefits must be responsible for creating the problem. Other conspiracy theorists need only a hunch and then embark on a search for evidence. And they find it. Everywhere. They refer to themselves as truth seekers. It is not the truth they seek. It is confirmation of their beliefs.

The conspiracy theory mindset is not harmless. Conspiracy theories have been linked to prejudice, war, and even genocide. The violent extremism of the Proud Boys is deeply rooted in conspiratorial thinking. HIV denialism by the government of South Africa, based on conspiratorial beliefs, caused the death of hundreds of thousands from AIDS in the early 2000s. The government of Zambia rejected food aid during a 2002 famine because of conspiracy theories about genetically modified foods. Many COVID-19 deaths can be tied directly to conspiratorial beliefs about the vaccine.

The primary defense against delusional conspiracy theories is to maintain an open society. Critical thinking requires freedom of thought and expression. While many people will continue to use conspiracy theories to protect themselves from painful truths, balanced journalism and a robust public forum for debate are our best hope for a less delusional future.

26

THE MOTHER OF ALL CONSPIRACY THEORIES

We can't accept very comfortably that two nobodies, two nothings—Lee Harvey Oswald and Jack Ruby— were able to change the course of world history.
—HUGH AYNESWORTH

NO REPORT ON CONSPIRACY THEORIES is complete without a mention of the John F. Kennedy assassination. Kennedy was assassinated as he rode in a motorcade through downtown Dallas, Texas, on November 22, 1963. Police quickly apprehended Lee Harvey Oswald for the crime. Two days later, Oswald was shot and killed on live television by a local nightclub owner, Jack Ruby.

The Kennedy assassination was a gut punch for the nation. Six decades later, many people still remember exactly where they were when they learned the news of Kennedy's untimely death. Sadness, anger, and denial overwhelmed America and much of the world.

This singularly significant emotional event spawned oceans of conspiratorial thinking. For many, their sense of a rational, orderly world was upended. They quickly fell prey to wild and simplistic conspiracy theories, many of which did not answer the question of who did it and why. They only knew that Lee Harvey Oswald did not do it alone, and sinister forces must be afoot.

President Lyndon Johnson appointed some of the nation's most prominent and trusted figures, including Earl Warren, the Chief Justice of the Supreme Court, to a commission charged with documenting the facts and helping to calm the nation. The Warren Commission issued its comprehensive 888-page final report in September 1964. For many, this report only enlarged the pool of conspirators.

Countless articles, books, and movies have been produced on the subject. Many conspiracy theorists consider this event ground zero for their beliefs in unseen forces trying to control the world. The size and scope of a conspiracy like the Kennedy assassination make it a small step to other theories like a fake moon landing or a 9/11 hoax.

Whatever delusion feeds your deepest fears, there is undoubtedly a secret cabal behind it. And that cabal can almost certainly be tied back in some way to the Kennedy assassination.

27

SOVEREIGN CITIZEN MOVEMENT

He is not a citizen who is not disposed to respect the laws and to obey the civil magistrate; and he is certainly not a good citizen who does not wish to promote, by every means in his power, the welfare of the whole society of his fellow citizens.
—ADAM SMITH

FIVE-YEAR-OLD IRENE, recently taught the principles of bodily autonomy by her parents, was working contently on an art project when her preschool teacher attempted to redirect her to another classroom activity. Looking her teacher square in the eye, Irene responded with the question: "Am I not the boss of me?" Profound confusion about this question is at the heart of the Sovereign Citizen movement.

The FBI describes Sovereign Citizens as "anti-government extremists who believe that even though they physically reside in this country, they are separate or 'sovereign' from the United States." This conviction that "You are not the boss of me" leads to behavior ranging from nuisance court filings to fraudulent activity and even domestic terrorism.

The Sovereign Citizen movement stands out among conspiratorial beliefs as particularly delusional and harmful. Once a person accepts the idea that the usual rules don't apply, there seems to be no turning back. Like many conspiracy theories, there are many interrelated threads or rabbit holes that adherents may elect to explore, inflicting growing levels of harm on themselves and society.

Its roots can be traced back to the 1970s and such racist and antigovernment movements as Posse Comitatus and the John Birch Society. The movement gained momentum during the COVID-19 pandemic with an influx of people upset about government health mandates. The movement also proved attractive to QAnon and the January 6 insurrectionists because it provided a legal-sounding justification for engaging in illegal activities.

The underlying tenet of the Sovereign Citizen movement is that the government set up by our founding fathers was secretly replaced, either after the Civil War or in conjunction with eliminating the gold standard (you choose!). This conspiratorial false government then replaced common law (the body of law created by judges through written opinions) with admiralty law (the body of law governing navigation and shipping). Most importantly, the government then separated people into genuine and pseudo entities and created a "Treasury Account" that the false government could use to sell your future earnings to foreign investors.

Sovereign Citizens believe the issuance of birth certificates and Social Security numbers somehow manages to separate actual physical babies from their "corporate shell identity." They go to great lengths to distinguish between their sovereign "flesh and blood" identity and the "straw man" created by the false government. Conveniently, they believe that laws only apply to the straw man, not to actual "free" men and women.

The Moorish Nation branch of Sovereign Citizens claims that a nonexistent eighteenth-century treaty with Morocco grants them immunity from US law. Another branch believes that the Fourteenth

Amendment created a distinction between flesh and blood Sovereign Citizens and enslaved federal pseudo-citizens.

Sovereign Citizens view the law as a book of arcane spells, and lawyers and judges as wizards who are key to this massive conspiracy to enslave Americans. They interact with police, the government, and the courts using legal-sounding language to inoculate themselves from laws they would rather not obey. They are constantly searching for the perfect, magic combination of capital letters, punctuation marks, red ink, symbols, thumbprints, and pseudo-legal word salad to render judges and the police powerless. They genuinely believe that if they can come up with the right combination of words and symbols, police officers and judges will be forced to yield and pay billions in monetary damages.

Many Sovereign Citizens also have a warped fascination with the Universal Commercial Code (UCC), a comprehensive set of uniformly adopted state laws governing commercial transactions in the United States. They believe that by filing a UCC financing statement, they can tap into their hidden "Treasury Account" and issue bonds that can then be used to discharge their debts. Others have decided they have squatter's rights to abandoned or foreclosed properties. Many will create ersatz birth certificates, driver's licenses, and license plates. They believe gold fringes on American flags displayed in courtrooms prove that admiralty law is in effect.

Sovereign citizens are not just harmless cranks spewing pseudo-legal gibberish. The Southern Poverty Law Center describes them as "paper terrorists." When faced with the enforcement of laws they don't like, they respond with a blizzard of legal filings using the language of their movement. They burden the courts with frivolous, nonsense filings. They default on their debts. They don't pay taxes. They file phony property liens (sometimes for millions or even billions of dollars) that can damage individuals and businesses. They harass and intimidate judges, law enforcement professionals, and other public officials. The most extreme adherents turn to violence and domestic terrorism.

When you hear someone talking about straw man versus flesh and blood, or "free persons," or "I am not driving in a vehicle, I'm traveling in a personal conveyance," or "Here is my fee schedule. I take silver coins," know that they are confused about who is the boss of whom. The barbed hook of delusion has been set and is not easily removed.

28

HOLOCAUST DENIAL

Remembering is a necessary rebuke to those who say the Holocaust never happened or has been exaggerated.
—**UN SECRETARY-GENERAL KOFI ANNAN**

HOLOCAUST DENIAL can be defined as any attempt to negate the well-established facts of the Nazi genocide of European Jews. It takes many forms. Some deniers claim Jews were deported from the Third Reich but not exterminated. Others argue that 5 to 6 million Jewish deaths are a gross exaggeration. Others claim that documentary evidence was fabricated and that the whole thing was a hoax.

There is also a range of motivations. Neo-Nazis have calculated that promoting their movement requires discrediting the Holocaust. Arab deniers object to the portrayal of Jews as victims at the expense of Palestinians. Other deniers are simply racist or antisemitic and have settled on holocaust denial as a blunt instrument for expressing their hatred. Some Holocaust deniers and revisionists mask their antisemitism under the guise of free speech and cast themselves as academics and truth seekers. Holocaust denial has also manifested within the radical right, the white power movement, and the Ku Klux Klan.

The clash between First Amendment rights and violent hate speech on social media continues to play out. The United States Supreme Court has rebuffed attempts to criminalize hate speech, however odious it may be. In the United States tradition, the answer to bad speech is not censorship but more speech that counters lies with facts.

Some extremist figures such as Rep. Marjorie Taylor Greene of Georgia, former Fox News host Tucker Carlson, and Robert Kennedy Jr. have borrowed the imagery of the Holocaust, seeking to build a narrative of victimization around COVID-19 vaccines and mask mandates. Anti-COVID vaccination protestors even wore armbands of yellow stars to promote a false equivalence of comparing the genocide of 6 million Jewish people with the inconvenience of wearing a mask.

The social harmfulness of this unique combination of hate and delusion has been acknowledged in many countries as a threat to the social order and made punishable under the law. Holocaust denial is explicitly or implicitly illegal in 17 countries. On January 20, 2022, the UN General Assembly passed a resolution on combating Holocaust denial and distortion. This resolution calls on member states to call out and reject holocaust denial and take active measures so that no such depravity ever happens again.

It seems incredible that more than 75 years after the fact, people can still choose to deny the best-documented mass atrocity in human history. The evidence is overwhelming, yet the delusion persists.

29

UNIDENTIFIED FLYING OBJECTS

I am discounting reports of UFOs. Why would they appear only to cranks and weirdos?
—STEPHEN HAWKING

THROUGHOUT HUMAN HISTORY, objects in the skies above have defied explanation. These range from meteors and comets to planetary alignments or even simple figments of the imagination. Historically, they have often been treated as religious or supernatural omens.

The notion that the most obvious explanation was intelligent extraterrestrial life didn't capture the popular imagination until the dramatic advances in aviation technology that accompanied World War II. Before World War II, most unidentified flying objects were described as cigar-shaped. In 1947, a civilian pilot named Kenneth Arnold reported seeing a series of flying saucers (not cigar-shaped) near Mount Rainier in Washington State. Reports of flying saucers soon became a daily occurrence in the U.S., completely supplanting cigar-shaped alien vessels. Another spike in UFO sightings began in the mid-1990s, coinciding with the development of U.S. Stealth aircraft.

The National UFO Reporting Centre (NUFORC) has compiled a dataset of global UFO sightings between 1910 and 2013 (more than 100 years). Most UFO sightings are in the United States (sorry, Asia, Africa, and South America—nothing to see there). Most UFO sightings within the U.S. are in California and Florida (no surprise there). Conveniently, UFO sightings are a summertime seasonal event.

The Roswell Incident is the most famous UFO story and the most appealing to conspiracy theorists. In 1947, a military weather balloon crashed near Roswell, New Mexico. A farmer recovered debris from the crash, which the U.S. military then quickly claimed. The incident languished in relative obscurity until the 1970s when UFO enthusiasts began to develop and promote increasingly far-fetched conspiracy theories about alien spacecraft, alien remains, and military cover-ups. A series of documentaries were released, including "UFOs: Past, Present, and Future" and "UFOs: It Has Begun" narrated (of course) with all due seriousness by Rod Serling. United States Congressional inquiries, Airforce, and General Accounting Office investigations have thoroughly debunked these claims.

Another hotbed of UFO claims is Area 51, a highly classified U.S. Air Force facility in Nevada. The CIA and Air Force acquired the base in 1955 but did not acknowledge its existence until 2013. While it is reasonable for the United States Military to conduct aeronautic and weapons testing in secret, that intense secrecy has made the base ground zero for conspiracy theories and UFO folklore.

Bizarre theories about Area 51 abound, including reverse engineering of crashed alien spacecraft, collaboration with extraterrestrial beings, development of exotic energy weapons based on alien technology, time travel, teleportation, and even cloning alien viruses. Everything you can imagine relating to extraterrestrials (and a few things you can't) is apparently kept secret in Area 51.

An anonymous Facebook post in 2019 that proposed to "Storm Area 51" and "see them aliens" generated more than 2 million responses. Two nearby music festivals corresponding to the event attracted nearly three thousand participants. Fewer than 200 attendees actually trekked to Area 51. Seven were arrested. No aliens were seen.

In July 2023, a reasonably well-credentialed group of cranks and conspiracy theorists testified before the US Congress that the government knows much more about UAPs (Unidentified Anomalous Phenomena—the new preferred name for UFOs) than it is telling the public and has possession of both UAPs and "nonhuman" biological matter recovered from UAP crash sites. They provided no evidence to support these incredible assertions, only the certainty of their delusion.

30

FLAT EARTH

The church says the Earth is flat, but I have seen its shadow on the moon, and I have more confidence even in a shadow than in the church.
—FERDINAND MAGELLAN

GREEK PHILOSOPHER ARISTOTLE (384-322 BC) was among the first to argue that our planet was a round sphere rather than a flat disk. Before Aristotle, many people assumed the Earth was shaped like a disc or a plate. They imagined that if you sailed too far in one direction or another, you would fall off the edge. Thar be dragons.

Aristotle deduced that only a round sphere would create a circular shadow on the moon during a lunar eclipse. This astronomical observation was supported by another observation he made at sea while traveling to Egypt. He noticed that when a ship sails from the coast, it disappears gradually behind the horizon.

Airplanes, satellites, space shuttles, and other advances in modern science (not to mention ancient science) have confirmed Aristotle's conclusion about the Earth's shape. With governments, media, schools, scientists, and airlines allied against them, the only way Flat Earthers can sustain their beliefs is to embrace some vast and astonishing conspiracy theories.

Flat Earthers even disagree with each other. Some believe our planet is an infinite plane; others think it is a disc or a disc under a dome. Some have a religious or creationist basis for their beliefs. But they all have one thing in common: embracing a constellation of other conspiracy theories along with their faith in Flat Earth.

A reporter attending a recent Flat Earth convention observed very few booths or presentations on the Earth being flat. The attendees accepted a Flat Earth as fact, so discussing it seemed unnecessary. Instead, booth after booth and presentation after presentation focused on a raft of other conspiracy theories in a remarkable echo chamber of delusion.

Social media metrics and growing attendance at Flat Earth events indicate that the number of Flat Earth conspiracy theorists has spiked considerably over the last ten years. Starting in 2015, Flat Earthers Eric DuBay and Mark Sargent launched a series of conspiratorial-themed Flat Earth videos on YouTube. DuBay and Sargent became minor celebrities with hundreds of thousands of followers. Their videos have been viewed millions of times.

DuBay's video (and subsequent book) "200 Proofs Earth Is Not a Spinning Ball" is a cornucopia of nonsense. Proof number 1 states, "The horizon always appears perfectly flat 360 degrees around the observer regardless of altitude," and NASA photographs are fake. That is not proof; it is a claim utterly unsupported by evidence and easily refuted. Proof 29 asserts that we would notice the motion of Earth spinning eastward at 1,000 miles per hour and that the Earth can't be a sphere because sometimes the wind comes from the West! Proof number 113 asserts that in a spherical world, ships sailing and people standing on the other side of the planet would be upside down, which, of course, is impossible. DuBay presents specious arguments like these in rapid succession during public events for the illusion of fluency.

A vibrant community of Flat Earth "researchers" has developed on YouTube and other social media platforms. Search algorithms fuel the conspiratorial fire by normalizing these theories in a community of like-minded people. One of the more colorful characters was the daredevil "Mad" Mike Hughes. With financial support from the Flat Earth community, he launched himself in a homemade rocket, intending to take a picture proving that the Earth was a disc. Instead, he crashed and died in the California desert in February 2020.

Flat Earth beliefs on their own seem harmless enough. The problem is that these beliefs do not exist in isolation. They are just one element in an ecosystem of anti-scientific conspiracy theories. Proponents of a Flat Earth run for school boards. They spurn vaccines. They hurt themselves and others by medicating themselves with untested herbs and oils. They vote for demagogues and grifters who support their conspiratorial beliefs. They dismiss Aristotle in favor of DuBay. The underlying problem with delusional Flat Earthers is not one of geography or physics. It is one of trust.

31

HIV/AIDS DENIALISM

HIV does not make people dangerous to know, so you can shake their hands and give them a hug: Heaven knows they need it.
—PRINCESS DIANA

HIV/AIDS BURST UPON THE SCENE in the early 1980s as a frightening new disease that seemed to be targeting gay men and intravenous drug users. While the terms HIV and AIDS are often used interchangeably, HIV (Human Immunodeficiency Virus) refers to the virus itself, and AIDS (Acquired Immunodeficiency Syndrome) refers to the collection of infections and diseases that immunocompromised HIV carriers may experience as their illness progresses. In the epidemic's early years, an HIV diagnosis was considered a death sentence.

Because HIV was spread primarily by drug users exchanging

contaminated hypodermic needles and gay men engaging in unprotected sex, the initial global response to the disease included an extraordinary amount of hysteria and delusion. Religious communities viewed the disease as a vindication of their belief that God's wrath was coming down upon sinners. Conspiracy theorists had a field day imagining all sorts of unseen hands behind the spread of the disease.

The infection was first documented clinically in the United States in 1981 in a small group of intravenous drug users and gay men who contracted a rare opportunistic pneumonia infection. Not long after, an unusual cluster of rare cancers appeared in another group of gay men. The Centers for Disease Control and Prevention (CDC) formed a task force and began to monitor the outbreak. Early on, they called it Gay Related Immune Deficiency (GRID) but soon realized that term was misleading. Later, they called it the 4H Disease (heroin users, homosexuals, hemophiliacs, and Haitians). Cooler heads soon prevailed, and by late 1982, the disease became known as Acquired Immunodeficiency Syndrome (AIDS).

As with any new disease outbreak, understanding the cause, transmission, prevention, and treatment best practices only comes with time. Science happens in fits and starts, and incomplete information may be the only information available at any given time. The first report of a link between HIV and AIDS occurred in 1983, but scientific consensus on this link was not really settled for nearly a decade after that.

As research dollars poured into studying the disease, scientists began understanding how it was transmitted and what could be done to prevent infection. Therapies that would make the condition chronic instead of fatal took a very long time to develop, but needle exchange programs and the promotion of condom use quickly became the norm. The use of gloves and safe disposal of sharp instruments and bodily fluids that we see today in dental and medical environments are largely a byproduct of the HIV/AIDS epidemic.

The slow march of science and the fog of misinformation about the disease in the 1980s and 1990s led to discrimination, hysteria, and denialism, resulting in unnecessary illness and death. In 1984, thirteen-year-old Ryan White was diagnosed with AIDS. He contracted the disease from treatment relating to his hemophilia. He was denied school admission, and he and his parents were subjected to harassment and bullying. His youth and courage in the face of this deadly disease and this awful treatment made him a national figure for others struggling with the disease. Ryan died of complications from the illness in 1990.

Religious people and organizations responded to the disease based on their religious convictions about sexuality, sin, and morality. For religious believers convinced that the end times were just around the corner, an epidemic that killed sinners was a real blessing. They argued that HIV/AIDS was a punishment from God for sexual sin or drug use. Some believed that HIV/AIDS was a way for God to thin the population before the end times. Others thought it was a test of faith and that those who could overcome the disease would be rewarded in the afterlife.

This led to discrimination against people with HIV/AIDS and made it difficult for them to access treatment and care. As a result, people did not get tested, seek treatment, or even report their test results. Ultimately, this caused additional infections, suffering, and death.

The Catholic Church's response to HIV/AIDS in the 1980s and 1990s was mixed. It condemned the discrimination and stigmatization of people with the disease and called for increased research and access to treatment. But at the same time, it taught that HIV/AIDS was a punishment from God for sin. The Church also stood by its long-time stance against using condoms, even in high-risk situations, and promoted abstinence as the only sure way to prevent HIV transmission.

Conspiracy theories sprout from a crisis like weeds in a garden. The HIV/AIDS epidemic was a fertile environment for conspiratorial nonsense. The low-hanging fruit for conspiracy theorists was that the whole thing was a hoax created by hidden hands for perverse purposes. A perennial favorite invisible hand is the CIA, producing and releasing HIV/AIDS as a biological weapon for unknown but undoubtedly nefarious reasons.

And if you ask the conspiracy theorists' favorite question, "Who gains?", then—of course—the pharmaceutical industry created the disease to sell more drugs. And we cannot forget the alternative medicine crowd with their herbs, oils, and pressure points trying to make a buck or two from the crisis.

The first case of HIV/AIDS was reported in China in 1985, but the Chinese government did not even acknowledge the epidemic until 1989. As a result, the epidemic snowballed, and by 2000, an estimated 1 million people were living with HIV/AIDS in China. Today, the Chinese have a model program called "Four Frees and One Care," which provides free HIV testing, counseling, prevention, education, antiretroviral therapy, and care for people living with HIV/AIDS.

The deadliest example of governmental HIV/AIDS denialism was in South Africa from 1999 to 2005. Thabo Mbeki succeeded Nelson Mandela as President of South Africa on June 14, 1999. He refused to believe that HIV caused AIDS and appointed a Presidential Advisory Panel that supported his beliefs. He blocked the use of antiretroviral drugs for AIDS patients and promoted herbal remedies such as garlic, lemon juice, and beetroot. His Health Minister soon acquired the nickname "Dr. Beetroot." Mbeki's delusions have been blamed for the preventable deaths of 350,000 people from AIDS.

A website designed to fight AIDS denialism (www.aidstruth.org) was created in 2006 and operated until 2015 (they consider their work done). This site provided links to the best available science associated with HIV/AIDS. As part of their effort to convince the dwindling pool of remaining denialists, their website also included a detailed listing of two dozen AIDS denialists who ultimately perished with infectious illnesses characteristic of AIDS. Most of these sustained their denialist fantasy right up to the end.

Worldwide, HIV/AIDS is estimated to have caused over 40 million deaths. Many of those deaths can be directly attributed to the slow and inadequate response by political leaders and governments. In sub-Saharan Africa, there was a widespread belief that AIDS could be cured by sex with a virgin. This myth is likely based on wishful thinking that the blood of a virgin is pure and can cleanse the body of disease. This delusion is doubly tragic, as the probability of transmitting the virus to innocent women is high, and the possibility of curing it is zero.

HIV today is considered a chronic condition and is no longer a death sentence. The stigma of having the disease is a fraction of what it once was. Some treatment protocols can reduce the viral load in HIV patients to undetectable levels. While some still cling to the denialist delusion, they remain on the extremist fringes and retain no hold on the reins of power.

32

CLIMATE CHANGE DENIALISM

I believe that global warming is a myth. And so, therefore, I have no conscience problems at all and I'm going to buy a Suburban next time.
—JERRY FALWELL

CLIMATE CHANGE DENIALISTS express doubt about or dismiss the scientific consensus that the Earth's climate is warming because of human activity. This group comprises a strange brew of fossil fuel industry self-preservationists, conspiracy theorists, political conservatives, and people who choose to wallow in delusion simply because they don't want to accept the behavioral change and cost implications of a warming planet.

The fossil fuel industry recognized in the 1970s that global warming could portend a shift away from carbon-based fuels and threaten its business model. For decades since, it has funded disinformation campaigns.

Their most significant win was the 2000 election of former Texas oilman George W. Bush to the U.S. presidency. The industry's campaign of climate change denial soon became presidential policy, and climate change denialism has skewed politically right of center ever since. According to the New York Times, from 2008 to 2017, the Republican Party went from debating how to combat human-caused climate change to arguing that it does not exist. During the 2016 presidential election cycle, every Republican candidate questioned or denied climate change.

Climate change denialism gives the appearance of a legitimate debate where there is none. Denialists allege a global warming conspiracy, cherry-pick contradicting data or studies, and claim that any statistical uncertainty invalidates the data while rejecting probabilities and mathematical models. However, the evidence supporting human-caused global warming is both stark and overwhelming. A 2021 review of a random sample of 3,000 academic climate change studies found only four skeptical of human-caused global warming. That is an extraordinary level of consensus.

Even without science, the natural world around us speaks volumes. Polar explorer Will Steger has testified before the U.S. Congress as an eyewitness to the ongoing catastrophic consequences of global warming in over 45 years of polar expeditions. In 2021, a massive chunk of ice broke off Antarctica, forming an iceberg the size of Rhode Island. Over 80% of the glacier atop Mount Kilimanjaro in Africa has melted.

But climate change denialists are quick to obsess about one data point, such as a cold snap, ignoring compelling trends in the opposite direction. They point out the effect of sunspots and overlook the fact that CO_2 emissions are vastly more impactful. They note that climate has constantly been changing, not realizing that past changes were often planetwide disasters with massive extinction events that we may want to avoid repeating.

Like many delusions, climate change denialism is based more on emotion than science. Unfortunately, the strident self-righteousness of activists like Greta Thunberg causes denialists to double down on their rejection of science.

As our climate warms, there will be winners and losers in different parts of the world. We will have to adjust to those changes as best as we can. That adjustment will be easier if we accept and act on the best science available and let go of delusional thinking about how we wish things were.

33

CORONAVIRUS CONSPIRACY THEORIES

People say: "Oh, Covid only affects people with pre-existing health conditions," like that's alright.
—ROSIE JONES

AS COVID-19 BEGAN TO SPREAD aggressively in early 2020, another viral phenomenon was right on its heels. Coronavirus misinformation and conspiracy theories exploded on social media platforms around the world. The World Health Organization warned in 2020, "Fake news spreads faster and more easily than this virus and is just as dangerous."

One theory was that 5G networks somehow contributed to the spread and severity. 5G is simply the latest evolution of wireless communications technology that allows for faster communication by using different frequencies and multiple antennas. Most people don't understand 5G technology, so it seemed perfect for sinister actors to manipulate for fun and profit. Conspiracy theorists observed that the densest areas of COVID-19 infection correlated with the density of 5G towers. Never mind that, in both cases, that's where the people are.

5G/Coronavirus conspiracy theories quickly reached a massive audience, thanks partly to celebrities like Woody Harrelson sharing them with millions of followers on social media. However, this irresponsible nonsense had real-world repercussions, with over 75 cell phone towers being vandalized over two months in 2020 in Great Britain alone.

Another theory promoted by then-President Donald Trump was that Hydroxychloroquine was a safe and effective treatment for the disease. He called it "one of the biggest game changers in the history of medicine." It was not. A study published in the Biomedicine & Pharmacotherapy journal estimated that compassionate, off-label use of hydroxychloroquine during the first wave of COVID-19 wave was related to 16,990 deaths.

Another horrible idea was that ivermectin (a horse dewormer!) could be used to treat the disease. People who self-medicated with ivermectin became sick as a result. In Iran, hundreds died after drinking methanol, believing it would cure COVID-19.

Another stubborn conspiracy theory was that elites like Dr. Anthony Fauci, Director of the National Institute of Allergy and Infectious Diseases, and Microsoft co-founder Bill Gates intentionally spread the virus for power and profit. This theory resonated deeply with the anti-vax community despite having no basis in reality. The anti-vax community believed that the vaccine (or Fauci Ouchie) was riskier than the disease itself.

The grandest delusion of all related to wearing face masks to help slow the spread of the disease. Early in the pandemic, personal protective equipment such as N95 face masks was in short supply. The general public was advised not to wear masks so public health workers would have a sufficient supply.

That changed weeks later, as masks became more widely available, and epidemiologists better understood how the disease spread. That reversal from early guidance and because masking was mandated in many states resulted in high resistance. People who didn't want to be told what to do adopted the delusion that masks had no effect.

PANDEMIC DENIAL—THE HERMAN CAIN AWARD

Herman Cain was a prominent Republican businessman, former Chairman of the Federal Reserve Bank of Kansas City, and presidential candidate in 2012. During the early days of the COVID-19 pandemic, Cain publicly denied the severity of COVID-19 and spread COVID-19 misinformation. Days after pointedly not wearing a mask or social distancing at a rally for Trump's reelection campaign in 2020, he tested positive for COVID-19. He was hospitalized and died of the disease less than a month later.

Cain's name became synonymous with the irony of downplaying the disease that killed him, helped in part by a Herman Cain Award thread on the news site Reddit. Reddit users captured screenshots of social media posts by people who made public declarations of their anti-mask, anti-vax, or pandemic denial views. Their posts were marked "Nominated" if they were hospitalized with COVID-19 and "Awarded" when they died. Some COVID deniers recanted from their hospital beds and begged their friends and family to take precautions, while others carried their delusions to the grave.

34

YEAR 2000 BUG (Y2K)

Y2K is a crisis without precedent in human history.
—**EDMUND X. DEJESUS**, Year 2000 Survival Guide

Y2K is the biggest non-event of the millennium.
—**DANIEL MARTIN**, December 1999

THE DELUSION OF Y2K was the notion that the world would become unhinged from a computing flaw in forty-year-old software. Governments, businesses, and individuals bought into this fear and went to extraordinary lengths to prevent the problem and prepare for societal collapse.

In the early days of computing, storage and operating memory were scarce and expensive resources. Programmers had to craft programs and data into the smallest size possible. One technique for saving space was to abbreviate four-digit years as two digits. Computer programs could recognize "99" as "1999" but could not identify "00" as "2000." They might interpret "00" as 1900 or stop working altogether.

Alan Greenspan (Chairman of the Federal Reserve from 1987 to 2006), in testimony before the Senate Banking Committee in 1998, recalled early in his career being proud of his ability to squeeze space out of computer programs by omitting the first two digits of the year. It never occurred to him (or other programmers in the 1960s and 1970s) that those programs would last more than a couple of years, much less into the next millennium.

By the late 1980s, industry pundits began to predict that this Century Date Change problem (later called the Year 2000 Bug or Y2K for short) would lead to software and hardware failures in computers worldwide when the clocks struck midnight on January 1, 2000. Banking, insurance, utilities, military, and government systems using mainframe computers running legacy software were considered most at risk.

But lots of devices contain software and computer chips. What about elevators, air traffic control systems, nuclear power plants, appliances, and even automobiles? The fact that some computer software might fail quickly morphed into the fear that all computer software might fail. This could lead to a complete meltdown of civil society. Few people had the skills to modify or repair computer code, so the only action most people could take was to try to be ready for disaster when it struck. No one knew precisely what Armageddon would look like, but many were convinced they knew exactly when it would occur. Midnight on January 1, 2000.

Customers, regulators, and shareholders compelled businesses to prove that their systems and software would not fail. Governments created task forces, passed laws, and prepared contingency plans. Private companies worked feverishly to check for Y2K compliance and fix software as the deadline approached.

The January 18, 1999, Time magazine cover rang in the new year with the words "The End of the World!?!" Other publications issued "Y2K Checklists," advising people to have medicine, cash, and hard copy records of all financial transactions. Many families set up emergency bunkers with blankets, flashlights, and food supplies. Australia recalled its embassy staff from Russia over fears of what might happen if communications or transportation networks broke down. Televangelist Jerry Falwell's $28 videotape, "A Christian's Guide to the Millennium Bug," suggested that Y2K was a sign from God that the Rapture was at hand. He advised stocking up on food and ammunition. The term "Y2K" became synonymous with "Doomsday."

When the first day of the new millennium dawned, power plants continued humming. Automobiles remained blissfully unaware of what day it was. Airplanes, phones, and appliances still worked. Banks did not run out of money. Martial law was not declared. Bunkers sat empty. The new millennium had begun, and life went on as usual.

A couple of Y2K glitches did occur. Some computers displayed the year as "19100" instead of "2000" but continued to operate. Slot machines at a race track in Delaware stopped working. A non-critical alarm went off in a nuclear power plant in Japan and was quickly corrected. An Australian man was charged $90,000 for a video rental return because the computer calculated he had held the title for over 100 years. The Maine Bureau of Motor Vehicles issued "horseless carriage" certificates for model year 2000 vehicles based on the miscalculation that they were manufactured before 1916. An Italian man withdrew his life savings from the bank on December 31st as a precaution, only to have it taken at gunpoint later that evening. Y2K was a catastrophe for him, but not for the reason he feared.

An estimated $100 billion was spent in the United States (and over $300 billion worldwide) to upgrade software and hardware to be Y2K-compliant. This massive infusion of money and effort paid dividends from upgraded software and computer systems for years. However, the elevated level of investment ended as the year 2000 began, contributing to the bursting of the Dot-Com Bubble that followed shortly afterward.

Depending on your perspective, the continued viability of computerized systems was proof that the collective effort had succeeded. Or it was evidence of delusion, the biggest non-event of the millennium.

35

NESSIE, BIGFOOT, AND THE ABOMINABLE SNOWMAN

Believe in yourself, even if no one else will.
—BIGFOOT

BELIEF IN FABULOUS OR SUPERNATURAL ANIMALS is as old as humankind. The modern pseudoscience of cryptozoology celebrates every report of a sighting, conveniently ignoring evolutionary biology, the fossil record, and scholarly consensus. Cryptozoologists (also known as monster hunters) promote their delusional belief in undiscovered (and usually enormous) mystery animals with a combination of hoaxes, wishful thinking, and misidentification of mundane objects.

The Loch Ness Monster (or Nessie, as she is affectionately known) is an excellent example. The earliest written accounts of Nessie date back to seventh-century accounts of a sixth-century encounter with a "water beast" by Irish monk Saint Columba in the river Ness. While additional sightings were reported in the 1800s, it wasn't until the 1930s that Nessie became a worldwide phenomenon. In 1933, there were multiple sightings, news stories, and even a blurry photograph. The famous surgeon's photograph of 1934 was held up for decades as proof that the monster existed. Additional photos, film, and even footprints followed in the ensuing decades.

One of Nessie's most fascinating characteristics is that her appearance and physical characteristics vary wildly from observer to observer, seeming to correspond only to the imagination of the lucky witness. Nessie has been described as looking like a large stubby-legged animal, an enormous creature with the body of a whale, a salamander, a long-necked dinosaur, an otter or a seal, having a hump, or two humps or four humps, as jet black and 14 meters long, or 30 meters long, or 7 meters long in some recent drone footage. She has been observed on land leaving footprints with prey in her mouth and at sea making waves, wriggling and churning up the water, or disappearing in a boiling frenzy of bubbles.

Decades of so-called proofs have been discredited as deliberate hoaxes or dismissed as simply not credible. A plaster cast of Nessie's footprints turned out to have been planted using a hippopotamus-foot umbrella stand. Even the 1934 photo, the most compelling evidence of Nessie's existence for decades, wasn't debunked until the 1999 publication of "Nessie—the Surgeon's Photograph Exposed." The photograph turned out to be a picture of a toy submarine purchased at F. W. Woolworth, with a putty head and neck molded on top.

Cryptozoologists are nothing if not persistent. The 1934 Edward Mountain Expedition posted twenty men with binoculars around the loch for five weeks to no avail. A 1960s sonar study came up mainly with fish. The Loch Ness Investigation Bureau disbanded in 1972 after ten years of fruitless searches using 35 MM cameras with telescopic lenses. The Robert Rines studies of the 1970s and 2000s used sonar and underwater photography to produce images that, if you covered one eye and squinted with the other, showed something. Maybe.

More recently, the Loch Ness Center assembled a team of 200 volunteers using all the latest technology for a two-day comprehensive search in 2023. Observers streaming video from flying and underwater drones (including thermal drones) reported inexplicable but inconclusive activity. Researchers also reported mysterious underwater noises but were dismayed to learn that their recorder was not plugged in. Organizers later conceded that the distinctive "gloop" sounds they heard might have been, in fact, ducks.

Nothing will ever disprove the Loch Ness Monster for many believers. They pin their hopes for "something that explains everything" onto Nessie's broad shoulders, and she seems to be carrying them just fine.

Bigfoot (also known as Sasquatch) is another imaginary creature purported to inhabit old-growth forests, mainly in the Pacific Northwest of North America. Most indigenous cultures have some tradition of larger-than-life creatures inhabiting nether regions of the wilderness. But it wasn't until the late 1950s that Bigfoot began to capture America's imagination. A California logging crew reported giant, 16-inch human-like footprints in the mud near their camps. Tales of hairy, ape-like creatures and plaster casts of the giant footprints soon became national news. There have been more than 10,000 Bigfoot sightings nationally since the 1950s.

It wasn't until 2002 that the family of a deceased member of that logging crew discovered a cache of large, carved wooden feet stored in his basement. But by then, Bigfoot was firmly established in the American psyche.

Like Nessie, Bigfoot appears in various shapes and sizes. Some report him as six to nine feet tall. Others claim ten to fifteen feet. He may have black, brown, or reddish hair covering his body. Sometimes, his eyes glow yellow or red. Sometimes, his footprints show evidence of claws—other times, not. And sadly, there does not appear to be a female of the species.

Also, like Nessie, hoaxes, wishful thinking, and misidentification of mundane objects account for virtually all of the alleged evidence and sightings. The "Minnesota Iceman" on display at the "Museum of the Weird" in Austin, Texas, turns out to have been made of latex. The autopsy of a dead Bigfoot discovered in 2008 found an empty head, fake hair, and rubber feet. A viral 2022 Facebook claim turned out to be an April Fool's joke.

The Abominable Snowman (also known as Yeti) demonstrates that large mythical creatures are not just a European or American phenomenon. Yeti is a large (of course) ape-like creature sporting white or grey hair inhabiting much of Asia, especially the Himalayan mountain range. Yeti comes complete with anecdotal sightings, disputed photographs and videos, and plaster casts of giant footprints. Yeti hair samples taken from remote Buddhist monasteries have generally been found to be some type of bear, indicating that the capability of a bear to stand on two feet may be the source of many sightings.

The lack of credible evidence is no impediment for true believers who have bought into the ancient delusion of cryptozoology.

36

CHEMTRAILS

I'm not a conspiracy theorist—I'm a conspiracy analyst.
—GORE VIDAL

CHEMTRAILS is a conspiratorial adaptation of the word contrails. Contrails are the cloud-like vapor trails jet airplanes leave high in the sky. They dissipate in low humidity and are not visible from the ground. In high humidity, they persist and can be observed from the ground long after the plane has passed.

People of a conspiratorial bent falsely believe chemtrails are evidence of something sinister. They think governments or other evil forces routinely spray the planet with chemical or biological agents for nefarious purposes. Hence the name Chemtrails.

There is general agreement among conspiracy theorists that Chemtrails are deliberate and harmful. However, there is very little agreement on who is behind the conspiracy or which nefarious purpose is primary. Far-right groups favor the Chemtrail conspiracy theory because it fits well with their deep suspicion of the government.

A smorgasbord of potential nefarious purposes is available to cater to any deeply held fear, including weather, mind control, climate warming (or cooling!), testing or distributing pathogens, biological, chemical, or other superweapons, manipulating markets, or creating a new world order.

The simple idea of water vapor crystalizing in high humidity behind a jet airplane is a bridge too far for some. The delusion of sinister powers controlling the world from just out of view has much more appeal.

37

THE BIRTHER MOVEMENT

Was it a birth certificate? You tell me. Some people say that was not his birth certificate. Maybe it was, maybe it wasn't. I'm saying I don't know. Nobody knows.
—DONALD TRUMP

BARACK OBAMA WAS BORN IN HONOLULU, HAWAII, on August 4, 1961. His mother was from Wichita, Kansas, and his father was from Kenya. Obama spent his early elementary school years in Jakarta, Indonesia, and returned to Honolulu to live with his maternal grandparents in 1971. He has resided continuously in the United States since then.

An early biography misidentified Obama's birthplace, stating he was "born in Kenya and raised in Indonesia and Hawaii." This mistake would haunt Obama's candidacy for the United States presidency, which was launched in February 2007. Rumors began to spread on conservative websites that Obama was ineligible to be President because he was not a natural-born citizen, as required by Article Two of the United States Constitution.

Amid growing calls for proof of citizenship by conservative and Democratic primary opponents alike, in June 2008, the Obama campaign posted an image of his birth certificate scanned from the original typed and handwritten document, which was (and is) stored in a bound volume in a file cabinet on the first floor of the Hawaii Department of Health.

This would serve as compelling, legal, prima facie evidence of the fact of birth in any court proceeding in the United States. It should have stopped the Birther Movement in its tracks. But it did not.

Political campaigners saw the birther movement as a fundraising tool and a way to spread fear and doubt about their opponent. Racists were attracted to the movement because it reinforced their narrative. Conspiracy theorists found a deep vein of deception and duplicity to mine. And so, the movement grew.

Conspiracy theorists determined that the scan must be a forgery and that the Hawaii Department of Health was in on the scam. They demanded that Obama release the original document, not a scanned version. This was eventually done in 2011, mainly to relieve the Hawaii Department of Health from the burden of relentless, repeated requests for the original document.

The scope of the Birther movement grew to include false claims that Obama was a Muslim. These claims were debunked. A false claim was made that Obama's paternal grandmother witnessed the birth in Kenya. This was debunked. A Kenyan birth certificate for Obama turned up and was quickly revealed to be a forgery.

There were claims that Obama lost his citizenship by traveling to Indonesia. Or that he had dual U.S./U.K. or U.S./Kenyan citizenship. Or that he couldn't be a citizen because his father was foreign-born. Some claimed that his middle name was originally Muhammad rather than Hussein or that Barack Obama Sr. was not his biological father. All these claims were debunked.

Donald Trump became a vocal spokesman for the birther movement during his 2016 presidential campaign. Birther conspiracy theories became a lucrative source of direct mail and telemarketing fundraising for Trump and other conservative organizations. Many Trump supporters still believe Obama was foreign-born, Muslim, or both.

At the 2011 White House Correspondents' Dinner, Obama released his official birth video—a clip from the opening scene of "The Lion King," where Simba is anointed the future king and presented to the animals of the Pride Lands. Fortunately, that delusion did not catch on.

38

PSYCHIC SERVICES

I don't believe in astrology; I'm a Sagittarius, and we're skeptical.
—ARTHUR C. CLARKE

THE PSYCHIC SERVICES INDUSTRY plays on natural human curiosity and a desire to know more about the future. There is an astonishing amount of demand for this information, delusional though it may be. The psychic services industry employs over 90,000 people and generates more than $2 billion in revenue in the United States alone.

In 2003, skeptic Michael Shermer partnered with public television personality Bill Nye for the PBS Science Series "Eye on Nye" to test the effectiveness of psychic readings. Shermer conducted "cold readings" of five subjects using Tarot Cards, palm reading, astrological reading, psychic reading, and communication with the dead.

He prepared for this by spending one day with Ian Rowland's "The Full Facts Book of Cold Reading," which is a text that debunks claims of supernatural ability. Rowland provided probing set-up questions, high probability guesses, and stunning insights such as "You are wise in the ways of the world, a wisdom gained through hard experience rather than book learning." Shermer also found success with the oddly specific phrase, "I see a white car." Almost anyone can find a meaningful connection to a white car if their mind is right.

All five readings were considered a success, with one being the subject's best-ever connection with her dead father in ten years of trying. The fact that an amateur with one day of study could do this reasonably well shows just how vulnerable people can be to these very effective scams.

ASTROLOGY

> *Two things are infinite: the universe and human stupidity; and I'm not sure about the universe.*
> — ALBERT EINSTEIN

Astrology, a pseudoscience that traces its lineage back thousands of years, is the cornerstone of many fraudulent psychic services. Astrology concerns itself with the alignment of stars and planets, so it seems plausible enough that it might be related to or supported somehow by actual science (i.e., astronomy). It is not.

Modern-day practitioners claim to be able to describe personality traits and predict events based on birth dates and astrological observations. Astrology is often combined with other psychic services such as palm reading or Tarot Cards. These predictions and personality traits are generally presented as a horoscope.

Thousands of newspaper columns, magazines, and websites carry horoscope content based on the astrological calendar. Predictions are carefully crafted to be vague and subject to multiple possible interpretations. Frequently, they are not predictions at all but advice or general statements about the human condition that could apply to nearly anyone at any time. Some typical examples (lifted from a recent local newspaper) include:

- Mistakes teach you quickly.

- Things left to their own timing will unfold with grace.

- Don't judge yourself for your mistakes.

- Unconscious forces will rise to the surface of your awareness, and you will have a new understanding of them.

What the actual heck?

Estimates vary, but between 10 and 30 percent of Americans read their horoscopes regularly. Angry backlash from loyal devotees has humbled newspaper editors who have attempted to eliminate horoscope columns. Hundreds of astrology-related websites and phone applications are available for those seeking regular guidance from the ever-moving cosmos. Premium services are available for a fee if free services aren't enough to satisfy your thirst for knowledge of the future. And don't forget your lucky numbers.

FORTUNE TELLING

> *The history of the esoteric use of Tarot cards is an oscillation between the two poles of vulgar fortune-telling and high magic; though the fence between them may have collapsed in places.*
> **—MICHAEL DUMMETT**

Fortune telling conjures to mind a robed figure in a seedy back alley caressing a crystal ball for connections with the dead or visions of the future. However, fortune telling can also involve reading tea leaves in a cup, palm reading, spirit board (Ouija Board), tarot cards, and other fraudulent techniques. Fortune tellers will generally use whatever technique resonates with their client and claim that any predictions that emerge are visions or revelations from spirits.

Most psychic services rely (at least in part) on the antiquity fallacy, which posits that older ideas must be true because they have been around for a long time. If the antiquity fallacy is your thing, palm reading is as good a place to start as any. Palmistry spans many ancient and exotic cultures, including Sumeria, Arabia, Persia, India, and China. However, different cultures offer different and often conflicting interpretations of the various lines and features of the hand. In addition to the head, heart, and life lines that form the core of most readings, there are also astrological features such as the Mercury, Girdle of Venus and Fate Line, Jupiter, Saturn and Apollo mounts, Mars positive and negative, and so on. In practice, palm readers carefully examine the lines on your hand and then apply high-probability guessing, inferences from probing questions, and broadly generalized nonsense to predict your future.

Another fortune-telling technique with ancient and culturally diverse roots is tarot card reading. Tarot cards started as actual playing cards in 15th century Asia and Europe. By the 18th century, tarot decks began to be modified for divination purposes. Today, there are many different versions of tarot cards, each with slightly different intended uses. Tarot readers will formulate a question and draw cards for insight into the past, present, or future. Some use tarot card reading as a tool for introspection and spiritual growth. Others believe the cards magically provide answers or that supernatural forces determine which cards are dealt and what they mean. Careful consideration of clubs, diamonds, hearts, and spades might be equally effective.

Tea leaf reading originated in China, where tea was invented. Ancient Chinese tea drinkers noticed patterns and shapes left by tea leaves in their cups and determined that this could not possibly be random. The varied configurations of soggy tea leaves must mean something; with careful and imaginative study, they could divine those meanings. As tea made its way to Europe, tea reading went with it. Tea reading remains an inexpensive way to predict the future and is about as accurate as any other.

PARAPSYCHOLOGY

I was sued by a woman who claimed that she became pregnant because she watched me on television and I bent her contraceptive coil.
—URI GELLER

Parapsychology is the study of phenomena that mainstream science cannot explain. This includes telepathy, psychokinesis, astral projection, hypnosis, clairvoyance, reincarnation, extrasensory perception (ESP), and more. These phenomena provide the perfect mash-up of personal beliefs, careless research, and outright fraud to make them ground zero for delusional thinking.

Some parapsychology researchers attempt to use legitimate scientific research techniques to evaluate claims of supernatural or otherwise unexplained forces. However, these researchers have a fundamental evidence problem and have failed to advance the science in their fields of study. Instead, most practitioners shamelessly specialize in parlor tricks and fraud.

Telepathy consists of direct mind-to-mind communication that does not rely on any known human sensory channels. Telepathy appeals to parapsychologists because it is easy to fake results. Most apparently successful demonstrations of telepathy are later found to have used clever signaling techniques or fall apart when trying to replicate results.

Psychokinesis (also called telekinesis) is the control of physical objects by focused efforts of the mind using forces undetectable by scientific scrutiny. One of the most famous purveyors of this delusion was the self-proclaimed psychic Uri Geller. Geller was a magician and television personality famous for spoon bending and other parlor tricks conducted under the banner of the psychic power of telekinesis. All of Geller's tricks can be replicated using stage magic techniques.

Astral projection imagines various out-of-body experiences in which consciousness and the physical body are separate. Although many religious traditions (particularly ancient Eastern traditions) perceive the body and the spiritual self as distinct entities, attempts to demonstrate separation of consciousness from the body or soul-traveling on alternate planes of existence have yet to be validated.

The precursor to hypnosis was Mesmerism, made famous by Franz Mesmer in 18th-century Germany. Charles Mackay details the many magical claims made by the "Father of Modern Hypnosis." If you limit the definition of hypnosis to "reduced peripheral awareness and heightened suggestibility resulting from extreme relaxation," it does not rise to the level of delusion. But most medical claims attributed to the power of hypnosis do rise to that level. Smoking cessation using hypnosis has had mixed results, for example. Using hypnosis to recall childhood trauma is controversial (at best) in the scientific community and may even lead to the formation of false memories.

Reincarnation is the philosophical or religious concept of passing a person's spiritual essence from one body to another after death. This falls under religion rather than delusion, except to the extent that celebrities and parapsychologists promote it as part of their overall shtick. Claims of memory from past lives can easily be dismissed as fraud or delusion. Actress Shirley MacLaine, for example, claims to have had an affair with Charlemagne in a past life and visits from extraterrestrials on her front porch in Malibu. She has made no mention of any past life affairs with Joe Sixpack.

Clairvoyance is the ability to see the past, the future, or objects and events that are not otherwise discernable to the senses. Clairvoyants may also refer to themselves as psychics. One of America's most successful psychics was author and syndicated newspaper columnist Jeane Dixon. She predicted in a 1956 issue of Parade Magazine that the 1960 presidential election would be "dominated by labor and won by a Democrat" who would "be assassinated or die in office."

This apparent prediction of the JFK assassination made Jeane Dixon famous. Never mind her subsequent predictions that World War III would break out in 1958, that Richard Nixon would beat John F. Kennedy in 1960, that Russia would be the first to put a man on the moon, or that an apocalyptic war of Armageddon would occur in 2020.

Like many prophets and psychics, Dixon made many predictions, hoping that some would come true and she could ignore those that didn't. Dixon died in Washington D.C. at the age of 93 and was reported to have said on her deathbed: "I knew this would happen." She got that right.

Science has repeatedly demonstrated that psychic services have no validity or predictive powers. One would imagine psychic fraudsters would lose their appeal over time as ancient superstitions are supplanted by the scientific method, and the internet provides access to credible, evidence-based alternatives. But the desire for insight into ourselves and our future is strong enough for many to set logic aside. As a result, psychic services stubbornly refuse to fade into obscurity and irrelevance.

39

ALEX JONES AND THE SANDY HOOK SCHOOL SHOOTING

The majority of frogs in most areas of the United States are now gay.
—ALEX JONES

ALEX JONES is a far-right radio talk show host and holds the dubious distinction of being America's leading conspiracy theorist. Jones's news and merchandising website (Infowars) promotes conspiracy theories and serves as a sales platform for conspiratorial books and videos, survivalist merchandise, and ersatz health products such as brain pills and dietary supplements. Recent estimates put Infowar's revenue in the range of $50 million per year.

Jones cut his teeth in the conspiracy theory community after the siege of the Branch Davidian complex near Waco, Texas, in 1993. He saw only "malevolent forces" behind these events and hosted a call-in show on public access television to air his views. The bombing of the Oklahoma City federal building bombing two years later only added to his anti-government paranoia and hateful rhetoric. He directed and starred in a 1998 film called "America Destroyed by Design," which purports to show how U.S. sovereignty is being compromised for global interests such as the United Nations.

Like most conspiracy theorists, Jones does not limit himself to just one or two delusional beliefs. Jones points to unseen forces promoting a fake 1969 moon landing and the September 11 attacks. Jones is also a leading proponent of anti-vax misinformation and a full range of alt-right false and misleading news. Social media platforms such as Facebook, YouTube, Instagram, and X (formerly Twitter) have repeatedly banned Jones from their platforms due to violent, hateful, and discriminatory content.

Alex Jones is also a vocal gun rights advocate. His conspiratorial bent, combined with these views, has led him to make excuses for mass shootings in the United States, claiming that they are used to create guilt on the part of the average gun owner so they would accept having their Second Amendment liberties curtailed.

Alex Jones is probably best known outside the far-right political spectrum for his repeated, outrageous claims that the 2012 Sandy Hook Elementary School mass shooting was a giant hoax staged by fake actors. Jones went beyond assertions that the government was behind the attacks and accused family members, survivors, neighbors, and first responders of being part of the plot. The shooting victim's relatives sued Jones for trauma, threats, and harassment for knowingly promoting this lie.

Jones admitted knowing the mass shooting was real at trial, yet he remained defiant and unrepentant on his show. In 2022, Alex Jones and Infowars were found liable in defamation cases in Connecticut and Texas and were ordered to pay nearly $1.5 billion in damages. Jones has filed for personal bankruptcy and now spends his time trying to prevent his assets from being seized by the courts.

Many of Alex Jones's long-held, deep-seated conspiratorial beliefs are clearly delusional. His alt-right audience is delusional to the extent that they accept his unsubstantiated theories about deep-state conspiracies attempting to control their lives. But there are also elements of malice, greed, and manipulation on display. Hopefully, Alex Jones will serve as proof that conspiracy theorists with public platforms who target private individuals can face serious consequences in court.

40

CATFISHING

If you look for truth, you may find comfort in the end; if you look for comfort, you will not get either comfort or truth only soft soap and wishful thinking to begin, and in the end, despair.
—C. S. LEWIS

CHARLES MACKAY could never have imagined the uniquely modern and deceptive practice known as catfishing. Catfishing is pretending to be someone you are not, usually to a specific victim (as opposed to the world at large). The motives are typically sexual harassment or financial gain. Catfishing is a byproduct of the ease of anonymity in our connected world of social media and the internet. It can be particularly cruel to vulnerable people looking for friendship or a relationship online.

The term "catfishing" dates back to a 2010 documentary film called "Catfish" by Nev Schulman. Schulman cultivated a friendship over the internet with a 40-year-old housewife who was pretending to be an 8-year-old girl. The term gained popularity when University of Notre Dame football star Manti Te'o fell victim to a public catfishing hoax in 2013 by a man pretending to be a woman dying of leukemia.

The delusion of catfishing is unrelated to the duplicity of pretending to be someone else. The delusion lies with people who choose to believe what they want to believe in the face of scams that seem obvious to outsiders and even, in retrospect, to the victims themselves.

Creating a "throwaway account" with a false identity on most social media platforms is simple. Other people's photos and life facts can easily be appropriated so that the scammer appears as real as possible. Many social media users feel comfortable sharing personal information in online forums. This leads some to quickly form deep connections with others in their online community, making them vulnerable to predatory behavior, even if some of the information they post is more aspirational than true.

There are several catfishing warning signs. One giant red flag is a race to intimacy. Proclamations of love after a few days or weeks of contact might be too good to be true. Another warning sign is asking for money or a gift. Unwillingness to meet in person or talk on the phone should signal that something is not quite right. Brand new or inconsistent profiles across platforms are another warning sign, especially if they don't have many friends or the profiles look too good to be true.

The best defense against delusion is to trust your gut. If it is rushed or doesn't feel right, give it time and talk about it with someone you trust. There are other fish in the sea.

41

CONSPIRACY THEORY SATIRE — BIRDS AREN'T REAL

If it flies, it spies! Bird-watching goes both ways!
—PETER MCINDOE

IT IS A REAL CHALLENGE to create a satire of conspiracy theories because they can be so wild and outrageous to begin with. How do you out-crazy crazy? Peter McIndoe managed, however, when he created the Birds Aren't Real satirical conspiracy theory in 2017.

McIndoe's Birds Aren't Real movement claims that all birds in the United States were exterminated by the federal government and replaced with mechanical drones that are spying on us. Birds sit on power lines to recharge their batteries. They poop on cars as a way to track our movements. And, of course, any self-respecting conspiracy theory will trace its roots to the mother of all conspiracy theories. John F. Kennedy had to be eliminated because he was unwilling to kill all the birds.

Birds Aren't Real has become something of a social media sensation. The Birds Aren't Real Facebook group has over 100,000 members. McIndoe created a video titled "The Confession of Eugene Price," the story of a former CIA agent who came out of deep cover to explain his role in the early days of the conspiracy. This video received over ten million views on TikTok. McIndoe claims that more than a million people (who are in on the joke) call themselves bird truthers.

Real-life conspiracy theorists have responded to the satire, claiming that Birds Aren't Real is a CIA psychological operation put out there as a weapon against conspiracy theorists. Maybe you can't out-crazy the crazies after all.

RELIGIOUS DELUSIONS

42

MAINSTREAM RELIGION

In the past 10,000 years, humans have devised roughly 100,000 religions based on roughly 2,500 gods. So the only difference between myself and the believers is that I am skeptical of 2,500 gods whereas they are skeptical of 2,499 gods. We're only one God away from total agreement.
—MICHAEL SHERMER

THERE ARE MORE THAN 4,000 different organized religions (depending on how you define religion) worldwide today. Seventy-five percent of the world's population subscribe to some expression of Buddhism, Christianity, Hinduism, Islam, or Judaism.

Most of these 4,000 religions have two things in common. First, they hold some version of the Golden Rule as a core belief. That is, "Do unto others as you would have them do unto you." The second is the shared belief that the other 3,999 religions are fundamentally wrong in some very serious ways. Even Buddhism, known for its "big tent" acceptance of other gods and idols, is pretty sure that your notion of one true faith is a failure to recognize that a bunch of other faiths are at least as valid. Never mind the billion or so unenlightened people who don't subscribe to any organized religion at all.

Many believe that other sects within their religion are also fundamentally wrong in profound ways. Some of the most horrible religious violence occurs between denominations within religious groups. This is especially true when power dynamics combine with sectarian differences to create an awful stew of hate and violence. "The Troubles" between Catholics and Protestants in Northern Ireland are one example. Sunni and Shia conflicts in the Middle East are another. Ninety-five percent of their religious beliefs may overlap, but the five percent that don't are worth fighting and dying for.

Some would classify any religion as delusional, but I see religious faith as an entirely reasonable approach to understanding the nature of God and the meaning of life. People are entitled to subscribe to whatever beliefs they like.

Some religions do rise to the level of delusion, however. They either have outrageous origin stories, cult-like attributes (such as a highly controlling structure or a self-proclaimed Messiah), or beliefs or practices far enough outside the mainstream to qualify as delusional. The following pages describe a few of these.

43

TELEVANGELISM

Religion began when the first scoundrel met the first fool.
—VOLTAIRE

MOST CLERGY IN AMERICA earn humble salaries and work tirelessly to share their religious message. Christianity has always emphasized preaching the gospel to the world, and radio and television are simply tools to spread that message. Televangelism, as we know it today, has its roots in revival-tent preaching in the South, which later evolved into radio and television programs by popular itinerant traveling preachers.

Not all televangelists fall under the delusion umbrella. One of the early pioneers of televangelism was Billy Graham, who was widely admired and respected as a person of high integrity. Televangelism moves into the realm of delusion when its charismatic leaders fall in love with the money, fame, and power accompanying their profession.

Many celebrity televangelists have generated significant personal wealth from their ministries. They often preach a "prosperity gospel" that promises material, financial, physical, and spiritual success. This gospel underlies their aggressive fundraising tactics. Give me money now, and you will be rewarded a hundredfold. It also helps justify the mansions, yachts, luxury cars, and private jets by demonstrating that the prosperity gospel works, at least for them.

Televangelists can be relentless and shamelessly manipulative in their appeals for financial donations from their followers. For the most part, their organizations exist independently from established Christian denominations. Very few are members of the Evangelical Council for Financial Accountability. They are ultimately not accountable to anyone but themselves.

Oral Roberts was one of the first to promote what has come to be called the "prosperity gospel," a belief that God wants you to prosper in all areas of life, including financially and in health (through faith healing).

In the 1940s, Roberts founded the Oral Roberts Evangelistic Association and started conducting evangelistic and faith-healing drives worldwide. He began broadcasting his revivals on radio in 1947 and on television in 1954. Acting on a command from God, he founded Oral Roberts University in 1963 in Tulsa, Oklahoma. In 1977, he claimed to have had a vision from a 900-foot-tall Jesus who told him to build a hospital (later called the City of Faith). In 1983, Roberts said Jesus had appeared to him in person and commissioned him to find a cure for cancer.

In a 1987 televised fundraising drive, Roberts announced that unless he raised $8 million, God would "call him home." Impassioned tears accompanied this outrageous claim. Roberts raised over $9 million. At its peak, Roberts' organization had annual revenues of $120 million and employed thousands of people.

In 1988, the Oral Roberts University Board of Regents approved millions in endowment money to buy Roberts a Beverly Hills property to serve as his office and home, as well as a country club membership. Roberts reportedly loved fine Italian silk suits, diamond rings, and gold bracelets. These fineries would be airbrushed out of publicity pictures by his staff.

Jim and Tammy Faye Bakker were another example of excess. They started their careers with Pat Robertson's Christian Broadcasting Network in Virginia in 1966. Jim Bakker was the first host of a primetime talk show called "The 700 Club," a highly successful weekly Christian news/talk program on the Christian Broadcasting Network (CBN). Bakker and Robertson parted ways in 1972 over philosophical differences, and the Bakkers went on to help launch the Trinity Broadcasting Network (TBN) and start their own television ministry called The PTL Club.

PTL grew and was highly successful. The Bakkers built a headquarters complex called Heritage Village and a $200 million Christian theme park called Heritage USA in South Carolina. Success piled on top of success. The Bakkers lived a lavish lifestyle, complete with jewelry, luxury cars, mansions, and private planes. At one point, they brought in $1 million of viewer contributions weekly. When accusations of wrongdoing started to pile up in the late 1970s, Bakker doubled down and claimed he was being unfairly persecuted, leveraging the controversy to raise even more money.

Bakker was caught paying hush money to a woman he allegedly raped, Jessica Hahn, with $270,000 of ministry funds. Bakker stole millions from his ministry to fund his lavish lifestyle and was eventually indicted on multiple counts of mail fraud, wire fraud, and conspiracy. He was found guilty and sentenced to federal prison.

After serving five years, Bakker returned to televangelism in 2003 with "The Jim Bakker Show" out of Branson, Missouri. He abandoned his previous focus on prosperity theology and turned to apocalypticism and conspiracy theories. He claimed that Barack Obama caused Hurricane Matthew, that he had predicted the 9/11 attacks, and that colloidal silver supplements were a cure for the coronavirus. Conveniently, he sold these supplements on his TV show until the Attorneys General of New York, Arkansas, Missouri, the Federal Trade Commission, and the Food and Drug Administration got involved.

Another example of delusional excess is Kenneth Copeland. In 1967, he founded Kenneth Copeland Ministries and Eagle Mountain International Church in Texas. His "prosperity gospel" is broadcast worldwide on the Victory Channel. He has written that believers will get a hundredfold return on their investment by giving to God (through him).

Various estimates put his net worth anywhere from $300 million to $1 billion. He lives in a $6 million, 18,000 square foot mansion outside Fort Worth on an exclusive lake community. The ministry also owns multiple aircraft, including a $20 million Gulfstream jet. After purchasing the plane, he harangued his followers to donate millions more to upgrade his runway and hangar.

Copeland has been accused of using his jet for luxury personal vacations. He once stated that he did not want to fly commercially because he didn't "want to get into a tube with a bunch of demons." The IRS denied his 2009 application for tax-exempt status for the jet.

Early in the coronavirus pandemic, Copeland claimed to heal and protect his followers from the coronavirus by blowing the "wind of God" (his breath) towards the television. And oh, by the way, keep paying tithes if you lost your job in the economic crisis that the pandemic caused. If Copeland's money-grubbing doesn't move him into the realm of delusion, the "wind of God" certainly should.

Another example is Jimmy Swaggart, an American televangelist whose television ministry was established in 1975. In the 1980s, his broadcasts went from weekly to daily and expanded nationwide. By 1983, Swaggart's programs were broadcast on over 250 television stations, and in 1987, Jimmy Swaggart World Ministries and its Bible college earned $150 million in revenue, ninety percent of that from the TV ministry.

In 1988, Swaggart was accused of a sex scandal involving a prostitute. He later tearfully and publicly apologized, delivering what came to be known as his "I have sinned" speech on live television. He was suspended and defrocked by the Assemblies of God for this offense, resulting in the shrinking of his now non-affiliated, nondenominational ministry. Three years later, Swaggart was found in the company of a prostitute for a second time. This time, Swaggart explained to his congregation, "The Lord told me it's flat none of your business." They disagreed, however, and he was forced to step down again. Fool me twice?

44

FAITH HEALING

*Faith healers don't get jobs in hospitals for the same reason
that psychics don't win the lottery every week.*
—RICKY GERVAIS

FAITH HEALERS are the religious equivalent of a bad penny. No matter how often they are discredited and debunked, they keep coming back. And some may wonder if there is merit to it. After all, Jesus and the apostles healed the sick; why shouldn't we expect such miracles today?

Unfortunately, the overwhelming majority of faith healers use parlor tricks and emotional manipulation to create the appearance of miracles. Although faith-healing miracles are somewhat infrequent in the Christian Bible, faith-healing televangelists work this magic into every episode of their ministry.

Several Christian televangelists base their ministry on a concept known as the "Word of Faith." This is a peculiar mix of the "Prosperity Gospel" and faith healing, both powerful drivers of financial contributions.

One example is Benny Hinn. Best known for his miracle-healing crusades, usually held in large stadiums and televised worldwide, the Israeli-born preacher has a notable track record of unfulfilled prophecies, including the death of Fidel Castro in the 1990s (Castro died in 2016) and the end of the world in 1992 and again in 1999.

At his Miracle Crusades, Hinn has claimed to cure cerebral palsy, blindness, deafness, cancer, and HIV. Investigative journalists have repeatedly debunked these claims. A 2001 HBO documentary, "A Question of Miracles," investigated seven cases of Hinn's miracle healings and found that none were legitimate.

Peter Popoff is another egregious example. A German-born preacher who billed himself "The Miracle Boy Evangelist," Popoff claimed to have the power to heal the sick and foretell the future. Before revival events, his wife and staff would collect prayer cards and interview audience members about their health problems. During the event, Popoff would work the room and claim that divine revelation told him about specific illnesses of audience members while receiving this information in real-time from an accomplice through an earpiece.

Popoff's scam was exposed on the Johnny Carson show in 1986. Popoff went bankrupt shortly afterward, but he reinvented himself, and today, he is selling mail-order miracle spring water and debt erasers on television and the internet. Before changing his ministry from a for-profit business to a tax-exempt religious organization in 2006, he reportedly grossed over $20 million annually.

Other faith healers use similar techniques. With the aid of prayer cards, hidden cameras, and pre-show interviews by staff members, vulnerable miracle seekers are identified and targeted. Anyone with a visibly apparent physical condition is not allowed on stage. They are ignored or intercepted and returned to their seats.

Another common technique is to heal people of diseases they don't have. If the preacher convinces you that you are deathly ill and don't know it, it is easy to believe that a cure is available with just a little faith and a donation. The faith healer giveth disease and the faith healer taketh it away.

Faith healers regularly hire actors for wheelchair and crutch scams. In this scenario, everybody wins (except for people who are actually sick): the actor gets a gig, believers witness a miracle, and the preacher makes a buck. Win, win, win. Occasionally, homeless people are recruited, placed in wheelchairs, and healed so they can walk. If they show up a little drunk, they are simply touched by the spirit.

Some scams are so outrageous that it is hard to understand how anyone could fall for them. A technique called psychic surgery involves carefully masking the site of a pretend incision, reaching into a hidden bowl of chicken guts to remove the cancer, and then laying hands to heal the wound without so much as a scar. Hallelujah! Applause and donations, please.

What if a disaster happens, and people with legitimate problems get through the screening and land on stage? No problem at all. They are the exception that proves the rule. They simply did not have enough faith.

Christian Science is one of several Christian sects that favors prayer over medical science. It believes that illness is an illusion caused by faulty beliefs and that replacing those beliefs with healthy ones through prayer or mental concentration will cure the illness. While Christian Science does not consider its practices to be faith healing, it does believe that every Christian Scientist has the ability to heal. Christian Science Primary Instruction is a two-week course that will get you started (Way faster than medical school).

The Christian Science Sentinel weekly magazine is awash in miracle healing testimonials, citing relief from cancer, arthritis, deafness, multiple sclerosis, skin rashes, paralysis, vision problems, and more. They typically fail to cite an actual medical diagnosis.

Attempts to objectively study faith healing using clinical trials have yielded no data to support the practice. Data or no, believers are not likely to change their views. When believers choose faith healing instead of traditional medical care for serious illness, the outcome can be death or disability. A 1998 study of the deaths of 172 U.S. children after medical care was withheld on religious grounds found that 80% of the children would have had at least a 90% likelihood of survival with timely medical care. Their parents' delusion led to the death of these unfortunate children.

45

SERPENT HANDLING

Sorry, Homer. I was born a snake handler and I'll die a snake handler.
—**MOE SZYSLAK**, The Simpsons

SOME CHRISTIAN DENOMINATIONS practice venomous snake handling as a demonstration of faith and commitment. Practitioners frequently die from poisonous snake bites, so their churches remain small, but it is such a special kind of stupid that it bears examination as an extraordinary delusion.

Snake-handling Christian denominations cite multiple scriptural sources supporting this practice, including:

Luke 10:19—Behold, I give unto you power to tread on serpents and scorpions, and over all the power of the enemy: and nothing shall by any means hurt you.

Mark 16:17-18—And these signs shall follow them that believe: In my name shall they cast out devils; they shall speak with new tongues. They shall take up serpents; and if they drink any deadly thing, it shall not hurt them; they shall lay hands on the sick, and they shall recover.

Acts 28:1–6, which relates how Paul was bitten by a venomous snake and suffered no harm, convincing barbarians that he was a god.

Even biblical literalists can look at these passages and disagree about what believers are called to do. The power to tread on snakes doesn't mean you have to let them bite you.

Many trace the practice to George Went Hensley. He had been raised Baptist but became fixated on those few lines in the Bible about serpent handling. He ventured out into the wilderness to seek God's will, and lo and behold, he came upon a snake. He knelt over it in prayer and picked it up. It didn't bite him. This was all the evidence he needed to enter the ministry full-time and lead snake-handling revivals throughout Appalachia. You may not be surprised to learn that his string of successes ended in July of 1955 in Florida when a venomous snake bit him on the wrist. He refused to seek medical attention and died waiting for divine intervention.

Snake-handling churches today are notoriously publicity-shy. They tend to have small congregations in small towns and mistrust strangers. Some estimates put the number of congregations at roughly 100, mainly in Appalachia. Another estimate puts the number of snake-handling deaths in the United States at approximately one per year. This seems like a small number overall, but this is a lot in small communities.

Despite constitutional concerns about freedom of religion, most states have banned the practice. Some states only get involved when there is a fatality or limit enforcement to wildlife handling charges to skirt constitutional issues.

The mental gymnastics required to accept snake handling as proof of faith is extraordinary. For most Christians, "proving" one's faith in this manner violates the biblical command: 'Do not put the Lord your God to the test.' The scriptures tell them they can pick up snakes without being bitten, but they get bitten just the same. If they have enough faith, they won't get sick, but they do. When they die, it is no longer a test of faith. It was their time, and God was calling them home.

46

CATHOLIC PRIEST SEX ABUSE COVER-UP

As to marriage or celibacy, let a man take which course he will, he will be sure to repent.
— SOCRATES

CHILD SEXUAL ABUSE is not unique to priests or the Catholic Church. Abuse can occur wherever adults interact with children, including in education, industry, scouting, sports, and even within families. Cultural, social, financial, physical, and other power imbalances increase the risk and can result in severe and lifelong harm to victims—even suicide.

Other religions struggle with this issue as well. Many cults have sexual abuse by messianic leaders baked into their ideology. Sex abuse scandals and accusations of cover-ups have roiled the deeply conservative Southern Baptist Convention. Lutherans, Methodists, Mennonites, Mormons,

Orthodox Judaism, and Buddhists have also had to wrestle with this issue. In 2018, the Dalai Lama met in the Netherlands with victims of abuse by Buddhist monks and said that this was nothing new and that he was aware of allegations of abuse by monks going back to the 1990s.

Two factors make child sexual abuse by Catholic priests unique. The first is the vow of celibacy taken by priests, which not only selects a unique population of men but also impacts how they deal with normal human sexual desires as they live out their lives. The second is the degree to which the Catholic Church institutionalized abuse, cover-up, and the reshuffling of dangerous individuals over many decades with seeming impunity.

Since the fourth century, Catholic tradition has obligated priests to be as unencumbered as possible "for the sake of the kingdom of heaven" so they can devote their lives more freely to the service of God and humanity. To meet this obligation, they take a vow of celibacy when they are ordained, which requires them to abstain from indulging in sexual thoughts or behavior.

Most men who consider entering the Catholic priesthood struggle with celibacy. Some opt out precisely for this reason. They may want to have a family or view abstinence as not desirable, realistic, or achievable. Many Catholics venerate their priests precisely because of their commitment and devotion to their ministry. This gives priests power in relationships with their parishioners that may not exist in other environments.

Some priests accept the logic of Catholic Catechism that the personal sacrifice of celibacy is a gift from God that will help them focus on their ministry. But others are deeply conflicted by their sexuality (whether heterosexual, homosexual, or predisposed to sexual behaviors that could harm others) and accept celibacy to deny, defer, or otherwise deal with those conflicts. They may feel, for example, that due to societal disapproval, being abstinent as a priest may be a better option than being openly gay as a layperson.

Unfortunately, sexual desire is like physical hunger. If satisfied today, it will return tomorrow. Humans are fallible, and Catholic priests are no different. The vow of celibacy inevitably creates a population of priests who may be sexually frustrated and in positions of power over others. Given the intimate nature of the problem and the Catholic Church's teaching on sexuality in general and homosexuality in particular, there is no way to determine the full extent of the issue of sexually frustrated priests and the sexual abuse that may follow.

Two significant delusions arise from this awful stew of celibacy and institutional denial. The first delusion is on the part of the abusers themselves. These individuals have devoted their entire lives to the service of God and their fellow man. It takes a special kind of madness to decide that God won't mind that they use their position to deny their vows.

Even worse, abusive behavior, especially against children, is a massive betrayal that conflicts with virtually everything they profess to believe. Seriously delusional thinking is required for any priest to act in a way that will preclude his future in heaven.

The second significant delusion is on the part of the church hierarchy. By its very nature, the Catholic Church is inclined to believe in repentance, forgiveness, and redemption. Priests caught abusing are counseled, and if they are repentant, they are likely to be forgiven.

Adding to the difficulty faced by bishops is the simple fact that public and grievous sins by priests undermine the church's authority. This creates tremendous pressure to keep transgressions secret. In addition, there has long been a need for more priests. Once an abuser repents, the easiest course of action is putting them back to work.

U.S. bishops have knowingly transferred thousands of abusive priests to other parishes (and other countries) in positions where they could easily re-offend. They allowed the fear of scandal and their faith in redemption to trump the welfare of children.

Child sex abuse and the pressure to cover it up are not new. BishopAccountability.org publishes a digital archive of Catholic clergy credibly accused or convicted of sexually abusing minors. In the United States alone, from 1950 through 2018, over 7,000 priests (with more than 20,000 victims) were identified in this database. That is 5.9% of the total population of priests working in the U.S. during that period.

In the 1980s, accusations of abuse and cover-ups received significant media and public attention in the United States and worldwide. As claims became public, other victims would come forward with new claims against individuals publicly identified for the first time after being shuffled from one parish to another. Many claims were not made until years after the abuse occurred due to self-blame and shame on the part of the victims.

In 2002, the Boston Globe investigated allegations of sexual abuse by Catholic priests and cover-up by the Catholic Archdiocese of Boston. Their investigation and subsequent blockbuster reports led to the criminal prosecution of six Roman Catholic priests. Long-repressed victims had the courage to begin coming forward, and soon, there were hundreds of lawsuits and additional convictions. Awareness of the enormity of abuse and the emerging pattern of denial, cover-up, and redeployment spread well beyond Boston, becoming a global crisis for the Roman Catholic Church. The scandal uncovered by this investigation was made into the Academy Award-winning film "Spotlight" in 2015.

Like the priests who deluded themselves that God wouldn't mind, the bishops and others in the hierarchy of the Catholic Church chose to believe that criminal prosecution was not a viable path forward and that repentance and forgiveness were reliable enough partners to keep giving second chances to sexual abusers. But billions of dollars in damages awarded in civil lawsuits are helping the church warm to the idea that the redeployment of abusers may not be the best alternative.

Pope John Paul II apologized in 2001, calling sexual abuse by priests "a profound contradiction of the teaching and witness of Jesus Christ." Pope Benedict XVI also apologized and met with victims of abuse. Pope Francis apologized and, in 2019, convened a conference to discuss preventing sexual abuse by Catholic Church clergy. Large institutions move slowly, but the awful delusion of cover-up and redeployment may finally be lifting.

47

CHURCH OF JESUS CHRIST OF LATTER-DAY SAINTS (MORMONISM)

Take away the Book of Mormon and the revelations, and where is our religion? We have none.
—JOSEPH SMITH, JR.

THE CHURCH OF JESUS CHRIST OF LATTER-DAY Saints (LDS) is the fourth-largest Christian denomination in America. Significant differences with other Christian denominations include a major commitment to missionary work, an emphasis on tithing, and a robust social support network within the church. Polygamy was allowed in the early days of the church, but this practice has long been abandoned, except by some fundamentalist sects.

The principal theological difference between LDS and other mainstream Christian religions is that they believe that the original Church described in the New Testament had deteriorated and needed to be restored. They see themselves as that restored (latter-day) Church. The founding prophet Joseph Smith accomplished this revelation and subsequent restoration.

In 1820, in Palmyra, New York, Smith claimed to have a revelation from God. This revelation came partly on a set of golden plates from which Smith transcribed the Book of Mormon. Smith formally organized The Church of Jesus Christ of Latter-day Saints in 1830 with the publication of this now-sacred text.

Smith described the golden plates as thin, golden metallic pages engraved with "reformed Egyptian" hieroglyphics on both sides. Smith was directed to the plates by the angel Moroni. He eventually retrieved the plates from their burial place but was not allowed to show them to others. They could only see and lift the box holding the plates. Smith dictated the text of the plates not by reading the hieroglyphics but by looking through a seer stone in the bottom of his hat. The original plates were returned to the angel Moroni after transcription and were never seen again.

Joseph Smith experienced a great deal of persecution for his beliefs. He was besieged with lawsuits, poisoned, beaten, tarred, and imprisoned. Smith and his followers were driven from Ohio to Missouri, Missouri to Illinois, and Illinois to the Utah Territory. Some of that persecution was triggered by his belief in polygamy, and some simply by his claim to be a prophet. He died a martyr in 1844.

Despite its fanciful beginning, the church, which started with six members in 1830, grew to over 271,000 by 1900 and an estimated 17 million worldwide today. This is primarily the result of aggressive missionary work. Church members were promised great spiritual blessings, including eternal joy if they fulfilled their missionary duties and helped save souls.

The Church of Jesus Christ of Latter-day Saints has more than 70,000 full-time missionaries worldwide. These are typically young people between 18 and 25, serving in pairs, who have put aside school, work, and dating for two years to teach Mormon doctrine to those interested in learning more about it. This practice has the dual effect of growing membership and increasing the commitment level of young members of the Church.

Another distinction of the Church of Jesus Christ of Latter-day Saints is its emphasis on tithing. Tithing is donating a tenth part (ten percent) of your income to the church. Tithing has biblical roots and is not unique to the Church of Jesus Christ of Latter-day Saints.

While tithing in most Christian denominations is estimated at well under five percent of members, roughly eighty percent of Mormons tithe. This generates an estimated $8 billion per year for the church, which uses these funds to pay for facilities (temples, schools, etc.), missionary work, social welfare, and other programs.

The church also has a money problem. Their problem is too much money. One of the most closely guarded secrets of The Church of Jesus Christ of Latter-day Saints is how much money it actually has. A 2019 IRS whistleblower complaint filed by David Nielsen, a senior portfolio manager at Ensign Peak Advisors, valued its portfolio from just this one entity at over $100 billion. Nielson described the investments as a clandestine hedge fund, using false records to masquerade as a charity, stockpiling money, and misleading church members. Church leaders claim the funds are for continuing operations and the future but also admit that the church has significant resources.

The shunning process experienced by people leaving the church lends it a cult-like sheen. This has been such a big issue that the church has been forced to modify its language to make it seem less horrible. "Disfellowship" has become "formal membership restrictions," and "excommunication" is now "withdrawal of membership," which is still shunning in practice. Families are broken apart as members are instructed to break all contact with former believers. Any contact with doubters is limited. Mormons will tell you those things are cultural and not doctrinal, but it's still a genuine issue for former members.

The Church of Jesus Christ of Latter-day Saints is considered by many to be mainstream today. Still, there are enough cult-like elements, enough sleight of hand in financial operations, and enough magic in the origin story to classify it as delusional. If one were to declare himself a prophet today, there could be no better sacred text than one that appears on golden plates—only to the prophet—and then disappears when convenient.

48

JEHOVAH'S WITNESSES

The proselytizing fanatic strengthens his own faith by converting others. The creed whose legitimacy is most easily challenged is likely to develop the strongest proselytizing impulse.
—ERIC HOFFER

JEHOVAH'S WITNESSES are another self-described Christian denomination that deviates significantly from mainstream Christianity. Jehovah's Witnesses believe that Armageddon is imminent. They believe birthday and holiday celebrations (even Christmas and Easter) have pagan origins incompatible with Christianity. They do not venerate the cross or other traditional Christian icons or images. They are known for door-to-door preaching, distributing literature, refusing military service, and rejecting blood transfusions.

They also believe that while God will bring billions of humans back from death to a cleansed Earth after Armageddon, only 144,000 will be resurrected to life in heaven. This belief seems strangely at odds with their enthusiasm for recruiting new members.

Early leaders in the church predicted that 1914 would mark the end of a 2,520-year period called "the Gentile Times," when society would be replaced by the full establishment of God's kingdom on Earth (in other words, Armageddon). Jehovah's Witnesses subsequently predicted that the world would end in 1915, 1918, 1920, 1925, 1941, 1975, 1994, and 1997. The Governing Body of Jehovah's Witnesses hasn't picked the next date for Armageddon, but it is almost certainly just around the corner.

The evidence of delusion for this cult-like religion includes autocratic leadership, demand for sustained and total commitment from members, and intolerance for doctrinal dissent or independent thinking. Members who openly disagree with the group's teachings are expelled and shunned. They share a text with mainstream Christianity (the Bible), but they land in some really different places with it.

49

CHURCH OF ECKANKAR

*We use more discriminating intelligence when we
buy a used car than when we buy a religion.*
—DAVID CHRISTOPHER LANE

ECKANKAR IS A RELIGIOUS MOVEMENT founded in California in 1965. It is now headquartered on a 174-acre campus in Chanhassen, Minnesota. Founder Paul Twitchell described himself as the Living Eck Master (the highest state of God consciousness on Earth). High praise indeed. The group is estimated to have 50,000 adherents across 100 countries.

Eckankar's beliefs are mainly drawn from Sikh, Hindu, and other Eastern religious traditions. One of the basic tenets is that the Soul is separate from the physical body and can travel freely in different planes of reality. The group teaches simple spiritual exercises (such as chanting

"HU") to acknowledge the presence of the Holy Spirit.

Twitchell developed his spiritual philosophy with the help of his second wife, Gail Atkinson (whom he married when he was 54 and she was 21). When Twitchell died of heart disease in 1971, his widow claimed to have had a dream in which her husband appointed Darwin Gross as the new Living Eck Master. They later married (!), and he served as the leader of Eckankar for ten years. In 1981, he appointed Harold Klemp to succeed him.

Two years later, Klemp and Gross had an acrimonious split, which resulted in Gross's excommunication. Lawsuits were filed over claims of financial impropriety and intellectual property rights. When the dust settled, Gross left to start his own group and any traces of his involvement in Eckankar were thoroughly excised from its official history.

Controversy dogs the organization. Founder Paul Twitchell claimed (without evidence) he was taught Eckankar while traveling in Europe and India by a five-hundred-year-old Tibetan monk. It has also come to light that Twitchell's spiritual writings contain large chunks of material appropriated from sources he failed to reference. Harold Klemp gently describes Twitchell as a "master compiler" when, in fact, he was a shameless plagiarizer.

Plagiarism, five-hundred-year-old monks, and the enormous leap of faith required to accept the central tenet of soul travel push Eckankar into the realm of delusion. People will always want to believe in supernatural, unprovable things. They call themselves truth-seekers, even though they disregard any information conflicting with their beliefs. They build their lives around their faith and are often afraid to leave the community for fear of losing themselves.

50

DIANETICS AND THE CHURCH OF SCIENTOLOGY

Writing for a penny a word is ridiculous. If a man wants to make a million dollars, the best way would be to start his own religion.
—Attributed to **L. RON HUBBARD**

NO STUDY OF SCIENTOLOGY would be complete without exploring the fascinating life history of its founder—Lafayette Ronald Hubbard. L. Ron Hubbard was born in 1911 and began his career writing professionally for pulp fiction magazines. From 1938 through 1950, he was a regular contributor to the magazine Astounding Science Fiction. During World War II, he also had an undistinguished career as a U.S. Naval officer in the Philippines and Australia.

After the war, Hubbard became involved with an occultist group in California led by American rocket engineer Jack Parsons. Hubbard later broke from the group and eloped with Parsons' girlfriend, Betty. Much of the theoretical basis for early Scientology beliefs and practices seems to have come from this period of Hubbard's life.

In 1950, Hubbard published "Dianetics: The Modern Science of Mental Health," which spent 28 weeks on the New York Times bestseller list. Historian Hugh Urban described Dianetics as "arguably the first major book of do-it-yourself psychotherapy."

Dianetics was initially promoted as a new form of psychotherapy. The therapist (or auditor) used counseling to recall past traumatic events. The idea was to expose and erase painful memories (a process called clearing). The Dianetics movement grew swiftly, and small practices started appearing across the U.S. and the United Kingdom despite occasional arrests for practicing medicine without a license.

Hubbard also developed and patented a mechanical device called an E-Meter (electropsychometer) for use during the auditing process. The E-meter is a pseudoscientific electronic device for measuring and displaying electrical charges observed on the skin's surface.

Hubbard initially promoted Dianetics as a science. Hence, the importance of the scientific-looking E-Meter. His writings at that time were decidedly anti-religious. Unfortunately, the American Psychiatric Association and the American Medical Association failed to take Dianetics seriously. Medical journals of the day rejected Dianetics submissions for publication. The Food and Drug Administration (FDA) was also skeptical, regarding Dianetics as pseudo-medicine and pseudoscience.

The science of Dianetics wasn't selling very well to end users or regulatory authorities, and in 1951, the Hubbard Dianetic Research Foundation entered into voluntary bankruptcy. Dianetics intellectual property was purchased by supporter Don Purcell, who set up his own Dianetics center in Kansas.

At about the same time, Hubbard's practice evolved towards the metaphysical. He deduced that auditing could reveal evidence of past lives and inner souls. His second major book on Dianetics, "Science of Survival" (published in 1951), introduced these notions of past lives and spiritual life energy. Although Hubbard's development of the term "Scientology" was born of legal necessity, it also coincided with his migration away from science and into religion.

Hubbard eventually concluded that opposition by the scientific community and regulators could be reduced—and revenue opportunities considerably enhanced—by transforming Scientology into a religion. In December 1953, Hubbard formally incorporated the Church of Scientology in New Jersey. Hubbard planned to set up a chain of "Spiritual Guidance Centers" and charged customers for auditing services. Hubbard's new definition of Scientology was "the study and handling of the spirit in relation to itself, universes, and other life." In 1954, the first local Church of Scientology was set up in California.

The Church continued to grow through the late 1950s and early 1960s. Hubbard's concept of Scientology evolved rapidly during this time, even contradicting his previous teachings. Hubbard was highly charismatic, so true believers of that era relied on unquestioning faith in Hubbard more than in any particular doctrine. Even today, Scientologists are taught to read Hubbard's works (all considered sacred texts) precisely in the order they were written.

Despite millions of adherents worldwide, the Church of Scientology has only attained legal recognition as a religious institution in a handful of jurisdictions, including Australia, Italy, the United States (in 1993), and the United Kingdom (in 1996). Germany identified Scientology as an anti-constitutional sect. France considers the group a dangerous cult.

The Church of Scientology considers itself a religion but operates like a business. Member churches charge for auditing (averaging $500 per hour). Church-related courses are required to advance through the ranks of Scientology. Program costs increase as you go. These programs can run to tens or hundreds of thousands of dollars over time. Local churches are structured like franchise operations, paying the Church's management arm in Los Angeles a percentage of their gross income. The Church pays commissions to recruiters (called Field Staff Members) for new members who sign up for auditing or counseling.

Exposure to the full extent of Scientology beliefs depends on advancement through the auditing process and becoming increasingly "clear." This also corresponds to the level of commitment to the organization (and the level of spending). Advanced Scientology religious doctrine veers into the realm of science fiction. Auditing eventually leads followers to the buried trauma of the "Incident," featuring an alien being called Xenu. Xenu ruled a confederation of planets 70 million years ago and brought billions of aliens to Earth, only to kill them with thermonuclear bombs. Scientology doctrine's bizarre, mythological aspect is the subject of considerable ridicule.

The church has not been required to file U.S. tax returns since 1993 when it won tax-exempt status. Recent estimates put annual Church revenue from its many corporations, donations, and real estate holdings at more than $500 million.

The Church has often faced unfavorable publicity and heavy criticism of its beliefs and practices. In 1965, the Victoria, Australia, state government accused Scientology of brainwashing, blackmail, and extortion. It condemned Hubbard as being of doubtful sanity, having a persecution complex, and displaying strong indications of paranoid schizophrenia with delusions of grandeur. A French court convicted Hubbard in 1978 for obtaining money under false pretenses. The Church itself was convicted of fraud by France in 2009 (upheld in 2013).

Hubbard established an intelligence unit in 1966 to respond to critics. The Guardian's Office aggressively sought to discredit or destroy anyone regarded as anti-Scientology. Between 1966 and the early 1980s, it launched scores of covert infiltration, wiretapping, and document theft operations. Ultimately, the Guardian's Office only added to the negative publicity when senior Church officials were indicted and convicted in federal court of criminal conspiracy for obstructing justice, burglary, and theft of government property. Hubbard himself was named as an unindicted co-conspirator for his part in the operation.

Feeling that he had to keep moving to dodge transnational legal issues, Hubbard developed a small fleet of ships in the Mediterranean and North Atlantic called Sea Org. Scientologists around the world were encouraged to apply to join Hubbard in the good life aboard the fleet. Rather than glamor, many found significant dysfunction and controlling behavior from an imperious Commodore who was being attended hand and foot by young girls dressed in hot pants and halter tops. Thinking "discreditable thoughts" was a disqualifier for the high honor of serving, so members worked hard to keep their minds right despite the 100 hour weeks in squalid conditions with poor food and little sleep.

Hubbard's legal woes continued into the 1980s, causing him to live on the move at sea, by hiding out in an RV, or at his ranch in California. He continued to be involved in Church operations (although to a lesser degree) and to be paid handsomely for his work. Hubbard died in 1986 after suffering a stroke. After his death, church leaders told his followers that Hubbard's body had become an "impediment to his work" and that he had decided to "drop his body" to continue his research on another planet. Hubbard had indicated that he would be reincarnated, and return "not as a religious leader but as a political one."

A schism followed, with some followers forming rival churches (Ron's Org and Free Org) and others following one of Hubbard's influential Sea Org aides named David Miscavige. Today, Miscavige is considered the ecclesiastical leader of the Scientology religion.

Like the organization he leads, Miscavige has been dogged by lawsuits and criminal allegations, including claims of human trafficking, child abuse, forced labor, coercive fundraising, harassment of critics, and abuse of subordinates. David Miscavige's father, Ron, decided to leave (escape from) the Church and wrote a book in 2016 about his experience called "Ruthless: Scientology, My Son David Miscavige, and Me." The book details David's brutal approach to running the organization and the abuse and harassment that Ron Miscavige endured through "disconnection." Another Miscavige relative, niece Jenna Miscavige Hill, published "Beyond Belief: My Secret Life Inside Scientology and My Harrowing Escape" in 2012.

In recent years, the Church of Scientology has worked to improve its public image by investing in a 24-hour television network that highlights some of the church's positive community outreach (such as human rights activism, drug rehabilitation programs, and literacy programs) while omitting the controversial aspects of their beliefs and practices.

Close observers of Scientology can see an origin story of greed, deception, and fanciful thinking, as well as a pattern of controlling and criminal behavior at the hands of charismatic leaders. One does not have harrowing escapes from organizations that are not criminal, delusional, or both.

51

THE UNIFICATION CHURCH (MOONIES)

The time will come, without my seeking it, that my words will almost serve as law. If I ask a certain thing, it will be done. If I don't want something, it will not be done.
—SUN MYUNG MOON

THE UNIFICATION CHURCH was founded in 1954 by the Reverend Sun Myung Moon in South Korea. Like many modern religions, it started with a base of traditional Christian theology supplemented by Moon self-identifying as the next Messiah and offering a sacred text he wrote called "Divine Principle."

With an early emphasis on missionary work and recruiting, the organization quickly expanded. At its peak in the 1980s, the church claimed to have 3 million members, with large concentrations in Japan and the United States. Moon was a fervent anti-communist, which helped him develop relationships with important conservative political figures around the globe.

Travelers in the 1970s and 1980s may recall seeing strange-looking young people dressed in colorful robes and selling flowers. These were Moonies, church members who sought alms in airports and other public places where First Amendment rights allowed them to ask for donations.

The Unification Church also became known for mass weddings, called "Blessing Ceremonies." The practice often involved couples meeting each other for the first time. In 1982, 2,000 couples were married in Madison Square Garden, and in 1988, thousands of Korean members were paired with Japanese members to promote unity between the two nations. In 1997, a Blessing Ceremony was held in Washington, D.C., for 28,000 couples. While this garnered a lot of publicity, some of these couples were already married, and others legally married later in a separate ceremony according to the laws of their home countries.

Another source of controversy for the Unification Church is its history of "spiritual sales." This practice involves scanning obituaries and reaching out to bereaved relatives to sell goods or services at exorbitant prices, promising supernatural benefits such as preventing bad luck or easing the souls of loved ones into heaven. The 2022 slaying of former Japanese Prime Minister Shinzo Abe was triggered by a grudge against Abe for supporting the Unification Church. The assassin's mother spent $1 million with the church in a spiritual sales scam, shattering the family's finances.

Defectors accuse the organization of being a cult. Disaffected members describe becoming wholly absorbed in the organization, breaking ties with their families, and raising money for the church. They accuse the church of deceptive recruiting methods, abusive practices, and manipulative financial techniques. In the 1970s and 1980s, hundreds of members were abducted by their family members and forced to undergo involuntary deprogramming, according to The New York Times 2012 obituary for Moon.

Any religion with a charismatic, self-proclaimed messiah is at risk of being labeled delusional. Add in spiritual sales, mass weddings, and cult-like manipulative behavior, and the delusion becomes clear.

52

END TIMES—THE MAYAN APOCALYPSE

When the world comes to an end, I want to be in Cincinnati. Everything comes there ten years later.
—MARK TWAIN

THE MAYAN EMPIRE STRETCHED from southern Mexico across much of northern Central America for thousands of years before collapsing in the 8th and 9th centuries. The Mayans built vast cities, ornate temples, exquisite artifacts, and towering pyramids and developed a complex hieroglyphic writing system.

One of their outstanding accomplishments was the Mayan "long-count" calendar, which ran from August 11, 3114 BC (the date they believed the Earth was created) to December 21, 2012 (after thirteen cycles of 144,000 days). This calendar was even more precise than the

Gregorian calendar we use today.

Doomsday prophets and spiritualists projected new-age pop culture onto the Mayan calendar with the idea of a cataclysmic event as the calendar ticked down to "the end of time." Some thought there would be a magnetic realignment of the north and south poles or the arrival of aliens to enlighten or enslave us. At a minimum, there would be floods, earthquakes, tsunamis, and general destruction of the human race.

One of the more imaginative theories was that a mysterious planet named Nibiru was hiding behind the sun, waiting to emerge and collide with Earth. None of this happened, but the idea of a deadly collision with Nibiru (also known as Planet X) is still knocking around the internet as a theory of how the world might end.

The Mayan Apocalypse spawned hundreds of books and thousands of websites devoted to the idea that the world would end in 2012. Even Hollywood got in on the act with a campy science fiction film called "2012—We Were Warned."

From the Mayan perspective, the end of each 144,000-day cycle was a significant interval but not a destructive one. None of the rich hieroglyphic records kept by Mayan scribes foretold the end of the world. The historical records are mostly silent about what they thought about turning over into a new cycle. If anything, perhaps they viewed it much like we would view New Year's Day—a chance to be refreshed and start anew.

On that fateful day, thousands of mystics, hippies, and tourists brimming with delusion gathered at Chichen Itza, Mexico and other Mayan temples to celebrate Armageddon (or the awakening). Somehow, the cataclysm missed us. The sun continued to shine and Planet Nibiru remained as elusive as ever. As for spiritual rebirth, you be the judge.

53

END TIMES—HEAVEN'S GATE AND THE HALE-BOPP COMET

To every man is given the key to the gates of heaven.
The same key opens the gates of hell. And so it is with science.
—RICHARD FEYNMAN

AMATEUR ASTRONOMERS Alan Hale and Thomas Bopp coincidentally discovered a comet separately on July 23, 1995. This came to be known as the Hale–Bopp comet. Hale-Bopp was the brightest comet seen for many decades and was visible to the naked eye for 18 months starting in May 1996. It has a 2,399-year orbital period, so its next visit to Earth will be in 4325 (plus or minus a couple of years).

Heaven's Gate was a cult founded in 1974 by Bonnie Nettles and Marshall Applewhite (known as Ti and Do). Ti and Do identified themselves as related to Jesus and witnesses of Revelation. By the mid-1970s, they had attracted a following of several hundred people. They eventually stopped recruiting and instituted a monastic lifestyle.

They believed they could transform into immortal extraterrestrial beings and ascend to heaven. Initially, they thought a UFO would take them to "the next level." Later, they came to believe that the body was merely a container for the soul and that the transfer vehicle was less important. Like most cults, members were told that they would have to shed their worldly attachments to qualify for heaven. This meant not just possessions but family, friends, jobs, and individuality.

In 1996, the group rented a large home in Rancho Santa Fe, California. They also purchased fifty individual million-dollar alien abduction insurance policies, which seems odd for a group whose aim is to hitch a ride with aliens on a UFO to paradise. In early 1997, Marshall Applewhite determined that a spaceship trailing the Hale-Bopp comet was the vehicle that would take them to heaven.

Applewhite persuaded 38 followers to videotape a farewell message and prepare for ritual suicide. They wore identical black shirts and sweat pants, new black-and-white Nike athletic shoes, and "Heaven's Gate Away Team" armband patches. Members then took a fatal dose of phenobarbital washed down with vodka. They each had a five-dollar bill and three quarters in their pockets.

The 21 women and 18 men are believed to have died in groups of fifteen, fifteen, and nine over three days. Acting on an anonymous tip, the San Diego County Sheriff's Department discovered the bodies on March 26, 1997. All 39 bodies were ultimately cremated. Their alien abduction insurance policies were never redeemed.

The level of delusion required to believe you will get to heaven by spaceship is mighty high. The notion that you will somehow benefit in heaven from a million-dollar alien abduction insurance policy on Earth adds to the delusion. And the five dollars and seventy-five cents pocket money for the journey is just plain nuts.

54

JIM JONES AND THE JONESTOWN MASSACRE

Be careful of living your life based only on faith and signs, or you might find yourself standing in a South American jungle holding a glass of Kool-Aid.
—SHANNON L. ALDER

IN NOVEMBER 1978, a mass murder-suicide event in Jonestown, Guyana, killed 909 individuals. As incomprehensible as this seems, the power of a delusional and messianic leader over impressionable people cannot be underestimated.

Jim Jones was an American preacher and political activist who founded the Peoples Temple in Indianapolis in 1955. In 1965, he moved the operation to San Francisco and engaged in civil rights activism and in promoting a form of communism he called "Apostolic Socialism." At its peak, the Peoples Temple had over 3,000 members who followed a communal lifestyle. They turned over all their income and property to Jones, who became increasingly more controlling over all aspects of community life. Jones was particularly effective at recruitment in the African-American community.

Jones adopted the belief that people could become manifestations of God with supernatural gifts and superhuman abilities and that he was one of those people. This manifestation signaled the second coming of Christ and a millennial age of heaven on earth. He also believed that America was capitalist, racist, and fascist and that only socialism and communism were free of sin. He frequently warned his followers of an apocalyptic race war to come and that the only way to escape imminent catastrophe was to accept his teachings.

Jones also became increasingly paranoid over time and had visions of an imminent nuclear attack. After reading in Esquire magazine that South America was the safest place to escape nuclear war, Jones traveled to Guyana and Brazil to look for a place to relocate the Peoples Temple.

In 1974, the Peoples Temple purchased land in Guyana. Its members cleared fields, installed a power generation station, and built dormitories. The agricultural commune would become known as Jonestown.

The negative press for Jones began with stories of fraudulent faith-healing miracles and outrageous claims of clairvoyance. Disgruntled former members reported harsh treatment, public ridicule, and physical violence. Jones was accused of illicit drug use, requiring sexual favors from female church members, and raping male members. He increasingly used fear to manipulate his followers.

Jones saw the Guyana compound as a socialist paradise and a sanctuary from media scrutiny. Jonestown started in 1977 with 50 settlers and quickly grew to over 1000. Jones put them to work twelve-hour days farming, building the community, and studying his teachings.

By late 1977, several Peoples Temple defectors formed a "Concerned Relatives" group. Based on tales of starvation, torture, rape, and abuse, they petitioned State Department officials and members of Congress for assistance. California Congressman Leo Ryan pledged to get involved.

Fearful of a government raid on the commune, Jones began to speak of "revolutionary suicide" and staged defensive drills that he called "white nights" that included mock attacks and suicide drills.

On November 15, 1978, Congressman Ryan led a fact-finding mission to investigate the abuse allegations. After a series of tense meetings, Ryan gathered fifteen members who wanted to leave. Armed guards met them at the airfield and gunned down Ryan and four others. When Jones got word that his security guards failed to kill all of Ryan's party, he called the entire community to the central square and informed them that it was only a matter of time before U.S. military commandos came to kill them all.

The community was overworked, exhausted, sleep-deprived, and starving. They had been accosted day and night with threats and sermons over loudspeakers. And now it seemed they were under attack, just as they had been warned.

Church leaders handed out a drink mixture of Flavor Aid and cyanide. Children were to drink the concoction before their parents. Armed guards offered the choice of death by poison or death at the hand of a guard. While some thought the exercise might be another White Night loyalty test, most meekly waited their turn to die. Nine hundred and nine inhabitants of Jonestown—276 of them children—died on that day. Jones was found dead on the stage of the central pavilion with a self-inflicted gunshot wound to the head, a victim of his own delusion.

55

KENYA STARVATION CULT

With or without religion, you would have good people doing good things and evil people doing evil things. But for good people to do evil things, that takes religion.
—STEVEN WEINBERG

KENYA HAS A RICH CULTURAL HERITAGE and tolerance for diverse religious practices. Religious freedom is enshrined in the Kenyan constitution. More than 80 percent of the country identifies as Christian. The evangelical/born-again movement has proliferated in Kenya since the 1980s and has deeply influenced Kenyan politics and society. This movement is highly decentralized, often led by charismatic leaders with little government oversight.

The Kenya starvation cult murders (also known as the Shakahola Forest Massacre) were a modern-day mass murder/suicide event within a Christian doomsday cult called the Good News International Church led by taxi driver-turned-preacher Paul Nthenge Mackenzie. He practiced classic cult isolation and bullying techniques, urging members to dissociate from the rest of the world, leave their jobs, refuse medical treatment, destroy government documents, and not speak with anyone outside of the church if they wanted to go to heaven.

Mackenzie allegedly baptized his followers in ponds before telling them the world was ending and to starve themselves to death in order to "meet Jesus." Over 370 people (including many children) were found dead in early 2023, and hundreds more were declared missing.

Autopsies of the victims showed evidence of starvation, strangulation, suffocation, and injuries sustained from blunt objects. Local media outlets also reported cases of missing internal body organs and at least one person buried alive. According to one survivor, March 2023 was set aside for the children to die. The month of April was set aside for the women to die, and the month of May for the men.

The great tragedy of Pastor Mackenzie's murderous delusion is that it could have been stopped earlier. There were years of warning signs, including many complaints and several police investigations. In March 2023, Mackenzie was arrested and charged with telling parents to suffocate and starve their children. He was subsequently released after paying a bond of less than $100. The full extent of the tragedy became apparent just weeks later.

56

LIGHTHOUSE CULT

Here's an easy way to figure out if you're in a cult: If you're wondering whether you're in a cult, the answer is yes.
—STEPHEN COLBERT

ON THE SURFACE, Lighthouse International Group is a mentoring and life-coaching organization. Paul Waugh founded Lighthouse in 2012 in Great Britain. In 2022, the High Court in Great Britain forced it into insolvency for failing to make required legal filings or pay taxes and for willfully obstructing the court's investigation into its operation. It has since rebranded as the Lighthouse Global Community and added a veneer of Christianity to its coaching activities. Waugh claims to have worked with over 75,000 clients over 18 years.

People typically come to Lighthouse by joining a mentoring group.

They may be looking for a career change, help achieving a life goal, or simply a new direction. Initial coaching and mentoring are often positive and helpful. Lighthouse takes this further by identifying and targeting which of these ordinary, reasonably intelligent people have some need or vulnerability that can be exploited to manipulate, brainwash, and drain them of their financial resources through coercion, isolation, and bullying.

Waugh's coaching premise is that there are four levels of spiritual development. Level four is a conscious and present person, free of constraints and fear. Waugh is one of only a handful of people on Earth to have achieved Level 4. Level one (pretty much everyone else) is a chaotic, childlike state. Climbing from one level to another requires mentoring calls, time, effort, and investment.

Waugh teaches that most people are stuck at level one (chaos) mainly because of negative influences around them (i.e., family and friends). During mentoring calls, Lighthouse coaches probe into personal affairs to discover and exploit any dysfunction in the participants' family life. In a nutshell, Lighthouse teaches that anyone who does not want you to pour your money, time, and resources into the organization is toxic and dangerous. This is a classic cult isolation technique.

People who have left the organization describe how they were drawn in a bit at a time. Daily coaching calls went from minutes to hours long. The calls were recorded, and intimate personal details were delved into. Participants were also required to transcribe Paul Waugh's thoughts and ideas, a technique designed to gain commitment, monopolize their time, and isolate them from family and other outside influences. Initial coaching fees ramped up over time to as much as $100,000 for a year's worth of coaching.

The BBC did an in-depth piece on Lighthouse in 2023 called "A Very British Cult." It describes how people trying to leave the organization faced threats to undermine their careers or reveal intimate secrets recorded during coaching calls.

The company's website reflects a significant persecution complex and a deep level of paranoia. Negative social media posts or tweets are blamed on predatory and abusive trolls. The site actually keeps a count of the number of harassing tweets and posts it receives. An article on the site is titled "Why Are We Persecuted?"

Lighthouse Global views itself as a community, not a cult. Yet the key elements of delusion associated with cults are clearly present in this organization. They include a charismatic leader whose Level Four thoughts must be transcribed, the persecution complex, paranoia and sensitivity to criticism, isolation techniques, financial drain for members, and threats against people trying to leave.

57

BRANCH DAVIDIANS

If the Bible is true, then I'm Christ.
—DAVID KORESH

THE BRANCH DAVIDIANS were an apocalyptic religious movement initially founded in 1955 as a spinoff from the Davidian Seventh-Day Adventists established by Victor Houteff twenty years earlier. They purchased land just west of Waco, Texas, and built their headquarters at Mount Carmel Center. A young man named Vernon Howell (later known as David Koresh) joined the group in 1981. After tumultuous legal and leadership challenges (including a shootout with community members loyal to another leader), Koresh emerged in 1987 as prophet and President.

David Koresh identified himself with the Lamb, as mentioned in

Revelation 5:2 "And I saw a strong angel proclaiming with a loud voice, Who is worthy to open the book, and to loose the seals thereof?" He wanted to create a new lineage of world leaders. In practice, this meant he assumed complete control over church community members and could rape and abuse the children without challenge. This abuse, along with the stockpiling of weapons, eventually led to a siege by the Bureau of Alcohol, Tobacco, and Firearms (ATF) on February 28, 1993.

The Branch Davidians were warned of their approach when a local TV reporter who had been tipped off about the raid asked for directions from a mail carrier who happened to be Koresh's brother-in-law. Seventy ATF agents attempted to execute the search warrant and were met with a hail of gunfire.

Four ATF agents and six Branch Davidians died, and sixteen agents were wounded. The raid devolved into a standoff, and the FBI came in to take command. The FBI negotiated the release of 19 children. Those children confirmed the allegations of physical and sexual abuse by Koresh.

During the ordeal, Koresh engaged in "Bible babble" over the phone and threatened violence. He claimed not to be suicidal. He allowed some followers to leave in exchange for supplies. At one point, Koresh said that he would surrender if his sermon were broadcast on national radio, but then failed to do so. The FBI tried turning off the electricity, playing Tibetan chants over loudspeakers, and shining bright lights to disrupt sleep.

Convinced that Koresh would not surrender, the FBI began a final attack on the compound at 6 AM on April 19, 1993, the 51st day of the siege. Armored vehicles punched holes into walls and launched hundreds of tear-gas canisters into the compound. A fire broke out, and by the time the smoke had cleared, 76 of the 85 Branch Davidians holdouts (including 25 children, two pregnant women, and David Koresh) were dead.

Law enforcement learned important lessons from this event, including the need for communication and coordination between agencies. Conflict and lack of coordination between the ATF and the FBI led to confusion and mistakes, which may have ultimately contributed to the unnecessary deaths of Branch Davidians. They also learned the need for patience and a better understanding of the hostage-taker's motivations.

April 19th became shorthand among right-wing militia groups for government overreach and oppression. The Oklahoma City bombers selected that date for their attack on the Alfred P. Murrah Federal Building two years later.

Koresh had messianic delusions. His followers bought into those delusions and looked the other way when Koresh abused their children. The FBI certainly made mistakes in handling the crisis, but responsibility for this horrific massacre lies entirely with Koresh and his followers.

58

FALUN GONG

In the West, the spirit is separate from the body. In the East, these are things that are very real and concrete.
—LI HONGZHI

FALUN GONG is an offshoot of Buddhism founded by Li Hongzhi in China in the early 1990s and is now headquartered in New York. Members seek spiritual enlightenment through meditation, slow movement, and breathing exercises. On the surface, their religious philosophy emphasizes morality and the cultivation of virtue. Falun Gong is best known in the United States as a victim of religious persecution in China due to its leader's cult-like behavior, embrace of conspiracy theories, and overt opposition to the Chinese Communist Party.

Falun Gong was relatively unknown in the West until the Chinese

Communist Party declared in October 1999 that it was a "heretical organization" that threatened social stability. Falun Gong was excised from the Internet in China, and many of its practitioners were imprisoned and subjected to a wide range of human rights abuses, including forced labor, psychiatric abuse, and torture. Human rights groups believe the Chinese government imprisoned hundreds of thousands of Falun Gong practitioners, and thousands died in custody. Those who died may explain a sharp increase in supply for China's organ transplant industry.

Falun Gong does not have any of the usual trappings of a religious group, such as temples, churches, or religious rituals. It has very little organizational structure beyond the notion that spiritual authority is vested exclusively in the teachings of founder Li Hongzhi, who believes (among other things) that aliens walk the Earth, that race-mixing is part of an alien plan to overtake humanity, and that homosexuality makes one unworthy of being human. He teaches that practitioners can acquire supernatural skills such as telepathy and "divine sight" through moral cultivation, meditation, and exercises. He promotes a messianic view of himself as the only person who can save humankind from destruction.

Falun Gong operates several media companies, including the Shen Yun Performing Arts group that tours America each year, but Li Hongzhi's teachings are distributed principally over the Internet. These entities are considered far-right politically both in the U.S. and in Europe. They believe that Donald Trump is an angel from heaven. They don't believe in evolution. They promote QAnon's conspiracy theories and spread anti-vaccine misinformation.

Despite persecution by the Chinese government, it is estimated that the practice still has millions of followers in China and hundreds of thousands more across 70 other countries.

Falun Gong defines itself as a health-focused spiritual group concerned about human rights. Former members who have extricated themselves from the group describe it differently. But is Falun Gong a cult? While its teachings are apocalyptic, it does not display other common characteristics of a cult. Members can marry outside the group, have friends, and hold regular jobs. They do not live isolated from society. They are not asked to give the group their money or possessions. The focus is inward, aimed at cleansing oneself spiritually.

Yet Falun Gong is much more secretive, messianic, and controlling than many people realize. Li Hongzhi warns his followers that they face eternal damnation if they don't do what he says. He claims great supernatural powers and that only he understands the truth of the universe and can see humankind's past and future. He claims that being omnipresent means he can monitor their minds and behavior, thus creating fear and reverence and gaining absolute obedience. Whatever he says becomes infallible scripture.

Hongzhi teaches that his pupils don't get illnesses. But of course, they do, and when they become ill, they can only be cured if they have perfect belief in him and his teachings. One of Falun Gong's dirty little secrets is that practitioners frequently die from treatable medical conditions.

While practitioners are technically free to come and go as they please, social pressures and deep spiritual fears keep them engaged. Many have abandoned other social ties and face the daunting task of rebuilding their lives from scratch if they try to leave. Former members believe "Master Li" is exploiting a free labor force of exhausted zealots to amass power and influence.

Falun Gong is sincere in its disdain for the Chinese Communist Party. It spends millions disparaging the Chinese government on social media and through its media companies (New Tang Dynasty Television, the Epoch Times, and Sound of Hope radio station). There's an incredible demand for the comforting illusion that one big villain controls the world. Falun Gong provides this narrative for many.

59

INTELLIGENT DESIGN AND THE FLYING SPAGHETTI MONSTER

> *This case is not about the need to separate church and state; it's about the need to separate ignorant, scientifically illiterate people from the ranks of teachers.*
> —NEIL DEGRASSE TYSON

INTELLIGENT DESIGN is an offshoot of Creationism, a conservative Christian religious belief that the earth is 6,000 years old (despite fossil evidence to the contrary planted by God) and that living organisms were created in their current form from specific acts of divine creation as described in the Book of Genesis. Its central thesis is that the theory of evolution cannot explain the immense complexity of life on earth, which, therefore, must have been designed by a supernatural entity.

Intelligent Design is essentially a re-branding of creation science to replace evidence-based evolutionary science taught in schools with pseudoscientific beliefs consistent with a literal interpretation of the Christian Bible. A series of court decisions have ruled that Intelligent Design has not decoupled itself from its religious roots and thus cannot be taught in American public schools.

In 2005, the Kansas State Board of Education considered allowing Intelligent Design to be taught as an alternative to evolution in public school science classes. A concerned citizen named Bobby Henderson sent an open letter to the board arguing that his belief that a Flying Spaghetti Monster created the universe was equally valid as intelligent design and should receive equal time in science classrooms. After he received no response from the Kansas State Board of Education, Henderson published the letter on his website. "Pastafarianism" and its deity, the Flying Spaghetti Monster, quickly went viral as a parody of religion and a symbol of opposition to teaching intelligent design in public schools.

Henderson cleverly used the same logical fallacies and circular arguments as Creationists. The overwhelming scientific evidence pointing towards evolutionary processes is nothing but a coincidence because every time a scientist carbon dates an artifact, the Flying Spaghetti Monster is there to change the results with His Noodly Appendage.

Henderson used the striking inverse correlation between the number of pirates worldwide and global average temperatures over the last two centuries to promote pirates as "absolute divine beings" and demonstrate that correlation does not equal causation. Pastafarians celebrate International Talk Like a Pirate Day on September 19th.

Henderson once concisely stated that the dogma of the Church of the Flying Spaghetti Monster is the rejection of dogma. Millions if not thousands of devout, secretive followers support Pastafarian beliefs, which include:

- All evidence for evolution was planted by the Flying Spaghetti Monster to test the faith of Pastafarians

- Pastafarian Heaven includes a beer volcano and strippers

- Pastafarian Hell is the same, except the beer is stale, and the strippers have sexually transmitted diseases

- Pastafarian gospel encourages readers to try out the religion for thirty days, saying, "If you don't like us, your old religion will most likely take you back."

- Pasta strainers are officially recognized Pastafarianism religious headgear

Henderson's Flying Spaghetti Monster is ultimately an argument about the arbitrariness of teaching religious views as science in public classrooms since any one view is as plausible as any other. It is a masterpiece of absurdity, highlighting delusional attempts to replace logical conjecture based on overwhelming observable evidence with religious faith.

POLITICAL DELUSIONS

There is a cult of ignorance in the United States, and there has always been. The strain of anti-intellectualism has been a constant thread winding its way through our political and cultural life, nurtured by the false notion that democracy means that my ignorance is just as good as your knowledge.
—ISAAC ASIMOV

PEOPLE OF GOODWILL can differ enormously in their political beliefs. These beliefs are driven by education, cultural background, life experience, religious convictions, notions of fairness and morality, and so on.

Deep concerns about fairness will lead one person to sympathize with the fate of illegal immigrants, while another will decry the injustice of violating immigration laws. Socialists see unfairness in the unequal distribution of resources, while capitalists see unfairness in the unequal distribution of effort.

Political views are often firmly held and difficult to discuss calmly and rationally under the best conditions. Debate is even more difficult when there is disagreement about the underlying facts.

In this author's opinion, the political beliefs described below rise to the level of delusion. I hope those who disagree will appreciate that while we both wish to improve the human condition, we simply see different paths.

60

QANON, PROUD BOYS, JANUARY 6 AND THE TRUMP DELUSION

In a revolution, one triumphs or dies (if it is a true revolution).
—CHE GUEVARA

QANON

In October 2017, a person calling himself "Q" (Q Anonymous or QAnon) made his first post on the 4chan imageboard website. He claimed Hillary Clinton was about to be arrested. A few hours later, a second message said that Trump was planning to remove "criminal rogue elements." The post also contained cryptic references to George Soros, other Clinton staffers, and the CIA.

Q claimed he had access to high-level classified information involving the Trump administration and its opponents. Q soon began to promote the idea that there was a vast, global conspiracy of satanic, cannibalistic child sex traffickers conspiring against Trump. This claim appears to have roots in the Pizzagate conspiracy theory from just a year earlier.

Pizzagate occurred during the 2016 United States presidential election. Conspiracy theorists claimed that hacked emails from high-ranking Democratic Party officials contained coded messages linking the Comet Ping Pong pizzeria in Washington, D.C., with child sex traffickers. These unsubstantiated claims quickly went viral on social media outlets like 4chan, 8chan, Reddit, and Twitter. The unfortunate restaurant owner and staff were subjected to harassment, death threats, and even an arson attempt. At one point, they were assaulted by a true believer from North Carolina who fired an AR-15 in the restaurant while searching for evidence.

QAnon posts (also known as "drops") were deliberately cryptic, leaving fellow conspiracy theorists to tease out whatever meaning they could. In one sense, Q was quite an innovator among conspiracy theorists. He enthralled readers with clues rather than presenting claims directly. Followers searched for evidence that would confirm their beliefs. These "drops" are also known as "breadcrumbs." Those who analyze the breadcrumbs call themselves "bakers."

In addition to being cryptic, Q used a very conspiratorial tone, with phrases like "I've said too much," "Some things must remain classified to the very end," "Enjoy the show," and "Follow the White Rabbit!"

Q was a prolific poster until after Trump lost the 2020 election. His account went quiet for 18 months, with posts only resuming in June 2022. There is some debate about whether the new posts are from the original Q or an imposter with his account details.

The QAnon phenomenon is a "big tent" of wild predictions and conspiracy theories. While tending towards the right wing, QAnon's ideas and predictions were richly creative, including some of the following gems:

- People targeted by Trump would commit mass suicide on February 10, 2018.

- The CIA installed North Korean leader Kim Jong-un as a puppet ruler and that there would be "bombshell" North Korea revelations in May 2018.

- German Chancellor Angela Merkel is Adolf Hitler's granddaughter.

- Obama, Clinton, Soros, and others are planning a coup against Trump.

- The Rothschild family leads a satanic cult.

- An event called "the Storm" will occur where thousands of people will be arrested and sent to Guantanamo Bay prison. The U.S. military will then take over the country and the result will be salvation and utopia on earth.

- John F. Kennedy would appear alive in front of a crowd in Dallas on November 2, 2021, and announce Trump's reinstatement as President.

- QAnon influencer Austin Steinbart was actually sending back messages from his future self, using a time-traveling computer.

So many questions. So little time.

Posts and predictions like these went viral on far-right social media platforms and mainstream platforms like Facebook and Twitter. They were further amplified by retweets from Donald Trump and promoted by Russian and Chinese social media troll accounts.

The exact number of QAnon believers is unknown. Still, the group has maintained a significant online following despite efforts by social media companies like Facebook and X (formerly Twitter) to slow the spread of its dangerous ideas. QAnon followers figured prominently in the attack on the U.S. Capitol on January 6, 2021.

PROUD BOYS

The Proud Boys is a violent, right-wing American militia group. Its glorification of political violence has resulted in terrorist organization designations from Canada and New Zealand, and it has been banned from social media platforms Facebook, Instagram, X, and YouTube.

Best known for their opposition to left-wing groups (such as Antifa) and their support for former U.S. President Donald Trump, Proud Boys view themselves as white men under siege and espouse a white genocide conspiracy theory. They have adopted a range of other conspiracy theories (no one can have just one), including many offered by QAnon and the anti-vax movement. The number of members is unknown but is estimated to be between several hundred and several thousand.

In the 2020 presidential debates, Trump was goaded by moderator Chris Wallace to denounce white supremacists and right-wing militia. Trump's response, "Proud Boys, stand back and stand by," was taken as validation of their cause and became a rallying cry for the group.

JANUARY 6 ATTACK

After the Electoral College cast their votes for Biden in December 2020, Trump called for supporters to attend a rally before the January 6 Congressional certification of the election. He tweeted, "Big protest in D.C. on January 6th. Be there, will be wild!" Thousands of Trump supporters, including the Proud Boys and other right-wing militia groups, made the trip to Washington DC, intent on supporting Trump and overturning the election.

A series of rallies were held near the Capitol on January 5th and 6th, culminating with a "Save America" rally in the National Mall just south of the White House. Trump repeatedly told rally participants to march to the Capitol, urged them to "fight like hell," and even promised to be there with them.

The Proud Boys (along with Oath Keepers, Three Percenters, and other anti-government militia groups) marched to the Capitol grounds, stormed through the barricades, overwhelmed the police, and entered the Capitol. An estimated ten thousand rioters entered the Capitol grounds. Twelve hundred of them breached the building.

Many rioters wore riot gear and carried plastic handcuffs and weapons such as guns, knives, axes, chemical sprays, and stun guns. Pipe bombs were found outside both the DNC and RNC offices. They erected gallows in the Capitol Mall and chanted "Hang Mike Pence," "Stop the steal," and "Fight for Trump." Some carried American flags, Confederate flags, and even Nazi emblems. They caused extensive physical damage, repeatedly invoking Trump's messages as their call to action.

The House and Senate evacuated their chamber without certifying the election by mid-afternoon. Rioters roamed the chambers hunting for Nancy Pelosi, Mike Pence, and other imagined enemies.

One rioter was shot and killed by police during the attack. One hundred thirty-eight police officers were injured, and three later died of conditions relating to the attack. Four officers who responded to the attack committed suicide in the following months.

Later that evening, both the House and the Senate were called back into session, and after some debate, Congress voted to certify Biden's electoral college win at 3:24 in the morning.

The Proud Boys and these other militia groups undoubtedly viewed themselves as noble revolutionaries in the mold of Paul Revere or George Washington. But noble revolutionaries don't use force to override the voice of the people after losing an election and don't drag Nazi and Confederate flags through the United States Capitol.

Instead, they felt aggrieved and chose to believe what they wanted to believe. They immersed themselves in a delusional, self-reinforcing echo chamber of lies and conspiracy theories from dark corners of the internet. Even cable news network CNN was complicit in this echo chamber of lies. Fox News personalities Tucker Carlson and Sean Hannity later admitted they knowingly presented false information in an effort to retain television viewer engagement and prop up the company's stock price.

In the aftermath of the event, the U.S. House of Representatives impeached Donald Trump for incitement of insurrection, making him the only President in the history of the United States to be impeached twice. Trump was acquitted in a Senate trial held after he left office. Nearly 1,000 rioters were charged with offenses ranging from disorderly conduct to seditious conspiracy (including five Proud Boys). Hundreds have pled guilty or been convicted. Many of these have expressed remorse. Others still cling to their delusions.

THE TRUMP DELUSION

Steve Hassan's book "The Cult of Trump" makes the case that Trump's followers are not what people commonly think of as a religious cult. Yet his charismatic leadership style and vitriolic rhetoric resemble mind-control techniques and result in cult-like behaviors by his followers. Trump's techniques include:

- Repetition of words and phrases that leads the listener to think they must be coming from more than one source and are, therefore, trustworthy

- Extensive use of thought stoppers like "Fake news" and "Build the wall" and "Lock her up" to crowd out analytical thinking

- Gaslighting and outright lies

- "We versus They" mentality

- Authoritarian behaviors

- Intolerance of any questioning or deviation from the "playbook"

- Deification (many Trump supporters believe he was appointed by God)

Trump also cleverly mashes up a host of "others" so that his followers can pick and choose which-ever "other" devils them the most. His targets include women, LBGTQ+, Blacks, socialists, unions, immigrants (especially rapist drug dealers from Mexico), people from "shithole" countries, Muslims, environmentalists, fake news liberal journalists, and more. Weirdly, many of his followers fall into one or more of these groups.

In 2016, many Republicans supported Trump for President despite his public personality based on his policy positions (stopping illegal immigration, appointing conservative Supreme Court nominees, anti-globalism, renegotiating trade deals, getting tough with China, etc.). But his most committed supporters were more tied to personality than policy. They liked that when he was challenged or made a mistake, he would double down and not admit that he was wrong.

People who felt politically or economically powerless wanted someone to come into Washington and start upsetting some apple carts. Policy positions didn't matter nearly as much as political elites getting a punch in the nose. Many Trump supporters would have voted for Bernie Sanders (who is diametrically opposed to Trump on most policy issues) for the same reasons had he been on the ballot in 2016.

Trump may be occasionally incoherent, narcissistic, and self-absorbed, but there is little evidence that he is delusional. He remains "focused like a laser beam" on burnishing his ego, regaining power, and whatever else he believes serves his current needs. On the other hand, his followers must be willing to engage in an extraordinary suspension of disbelief to continue supporting him, given everything we know about the man and his destructive impact on political discourse and civil society. Thar be delusion.

61

DOMINION VOTING MACHINES

You've heard a lot over the past few days about the security of our electronic voting machines. This is a real issue, no matter who raises it or who tries to dismiss it out of hand as a conspiracy theory.
—TUCKER CARLSON (on Fox News)

The software shit is absurd.
—TUCKER CARLSON (private text message)

WHEN THE VOTES WERE COUNTED for the 2020 presidential election, and Joe Biden came out on top, runner-up Donald Trump and his supporters looked for ways to disrupt, delay, and deny the results. Claiming massive voter fraud, they pressured local, state, and federal officials to overturn the election and filed lawsuits across the country.

At the center of this whirlwind of lies were the claims that Dominion Voting Systems (the manufacturer of many voting machines used during the election) was owned by the "radical left" and that its machines switched millions of votes for Biden and against Trump. These claims were made without evidence. It is difficult to imagine how big a conspiracy would have to be to involve hundreds of thousands of machines that are widely distributed and physically controlled by independent county election officials nationwide.

MyPillow CEO and ardent Trump supporter Mike Lindell claimed to have "absolute proof" of interference by China and rigging of voting machines by Dominion. He went so far as to offer a $5 million reward in a "Prove Mike Wrong Challenge" for anyone who could show "packet captures" and other data he released were not valid data from the November 2020 election. Software engineer Robert Zeidman accepted the challenge and produced a report showing that Lindell's data does not "contain packet data of any kind and does not contain any information related to the November 2020 election." An independent arbitrator agreed. Legal fisticuffs have ensued.

Despite the lack of evidence and the improbability of the conspiracy, Fox News hosts and guests kept up a steady drumbeat of claims that Dominion's voting machines had been rigged to steal the election from Donald Trump. Dominion Voting Systems filed a $1.6 billion defamation lawsuit against Fox Corporation for promoting these false claims. Fox News settled for $787.5 million and acknowledged that they knowingly spread falsehoods about Dominion.

As Mike Lindell and Fox News have learned, delusion can be expensive.

62

SUICIDE BOMBING

Men never do evil so completely and cheerfully as when they do it from religious conviction.
—BLAISE PASCAL

ON SEPTEMBER 11, 2001, fifteen al-Qaeda terrorists hijacked four passenger jet airliners and perpetrated the worst attack on American soil since the 1941 attack on Pearl Harbor. Two thousand nine hundred ninety-six people were killed, and thousands more were injured.

After recovering from the initial shock of this audacious attack, Americans struggled to comprehend the level of hatred and delusion that would lead fifteen people to sacrifice their lives to hurt us in this way. What seemed stunning about the attack was that the point was not to achieve a military objective but rather to inflict pain and cause terror among civilians to achieve political goals. However, even this tactic was not new in 2001.

Suicide bombing as a tool of "stateless terrorists" was ushered into the modern era with the bombing of a U.S. Marine barracks in Lebanon by Shiite militants that killed 241 American soldiers in March of 1983. Less than a month later, another suicide bomber drove a truck full of explosives into the U.S. embassy in Beirut, killing 60 people. Islamic Jihad claimed responsibility for this attack. Initially, U.S. officials vowed no change in U.S. policy as a result of the attacks. However, President Reagan ordered the U.S. forces to return home less than a year later.

Suicide bombing is not uniquely Islamist, but this early apparent success, plus a combination of other political and religious factors in the Middle East, came together to help the tactic gain traction among Islamist organizations such as the PLO, Hamas, Fatah, Islamic Jihad, al-Qaeda, and the Al-Aqsa Martyrs Brigade. The number of suicide attacks continued to increase, and by 2002, more than 200 people were killed in 40 separate attacks. Action on Armed Violence (AOAV) recorded 1,191 suicide bombings globally between 2011 and 2015, resulting in 31,477 civilian deaths and injuries.

What kind of delusional thinking leads people to do this? The Quran explicitly forbids suicide (Quran 4:29), yet there is a seemingly endless supply of young men (mostly) who are willing to give up their lives to murder civilians.

Most of these bombers live in a society willing to accept such acts for the greater good. They are rarely lone fanatics but rather the product of recruiting organizations that promise they will be revered as martyrs. Their families receive money for their sacrifice, and the bombers are told they will receive blessings in heaven, such as harems of virgins. These incentives are a powerful motivator in impoverished and polygamous societies where a few wealthy men have many wives and many poor men have no wives. (It is a bit of a mystery how this deal helps the virgins, however.)

Other factors can include economic desperation, a sense of powerlessness, pride, and anger against an enemy seen as a cruel or brutal occupier.

SUICIDE BOMBING

Suicide bombing has been called the atomic weapon of the weak.

Suicide attacks are a cost-effective form of asymmetrical warfare. And they can have a powerful psychological impact. The attacker cannot be caught or punished because they died conducting the attack. Their commitment to the point of death results in fear and a feeling of hopelessness for the victims of their attacks.

When a young man seeks the heavenly rewards of donning a suicide vest, he explicitly closes the door on every other opportunity in life. He gives up the possibility that his economic condition can change. He gives up the chance of having a wife or a family. He gives up opportunities to work peacefully for change. He gives up the possibility that external forces could ever resolve the political or social issues that led to his tragic and fatal decision.

63

THE KIM DYNASTY OF NORTH KOREA

North Korea cannot change because its people don't realize that there is an alternative to their suffering.
— **PARK YEON-MI** (North Korean Defector)

KIM IL-SUNG EMERGED from the ashes of World War II as North Korea's founder and first leader. He came to rule North Korea in 1948 and launched the Korean War in 1950 to bring South Korea under his control. Kim built the Democratic People's Republic of Korea (DPRK) with the support of the Soviet Union based on Marxist-Leninist principles as a command economy with complete state control of industry, agriculture, education, and healthcare.

Over his nearly 46-year tenure as Great Leader, Kim became increasingly totalitarian. North Korea's founding principles of

Marxism-Leninism were subsumed into a uniquely North Korean ideology of "Juche," which is the idea that true socialism can only be achieved through economic self-reliance, strong defense, and an independent state. Having a solitary, absolute leader is central to the concept of Juche.

North Korea self-identifies (according to its constitution) as a democratic republic. In reality, it is a cult-like dictatorship, where elections are for show and feature only party-approved, single-candidate races. If you want to vote against pre-selected candidates, you must use a separate ballot box under the supervision of election officials. This is a one-way ticket to prison.

The human rights record of North Korea is considered by many to be the worst in the world. Hundreds of thousands are incarcerated in prison camps for political crimes and subjected to forced labor, physical abuse, and execution. Citizens cannot freely travel within the country, let alone abroad. The government controls all forms of communication. Accessing information from other nations carries severe penalties. Satellite images of North Korea at night show nearly complete darkness, starkly contrasting with the bright lights of neighbors South Korea and China.

Kim Il-sung expanded his grip on power by developing a personality cult. His mother became known as the "mother of Korea," his wife as the "mother of the revolution," and he himself as the Heavenly Leader. His place of birth has become a pilgrimage site.

The dynasty started by Kim Il-sung has morphed into a hereditary dictatorship, now in its third generation. This is North Korea's central delusion. After Kim Il-sung died in 1994, his role as supreme leader was passed on to his son, Kim Jong-il, and then to his grandson, Kim Jong-un, in 2011.

Kim Jong-il embarked on a process to deify the Kim dynasty. Billions were spent on ceremonies and monuments. Kim Jung-il became not just the son of Kim Il-sung but his reincarnation. His birth at sacred Mount Paekdu was attended by a double rainbow while local birds sang songs of praise (in Korean, of course). Never mind that his parents were in Stalin's war-torn Russia when he was born in 1942.

The Kim dynasty has presided over famine, floods, and the loss of its primary sponsor with the dissolution of the Soviet Union in 1991. Yet, its grip on power appears to remain absolute. The biggest challenge has been to find ways to sustain North Korea's economy while staying isolated enough to maintain political control. A shortage of arable land in North Korea has forced it to rely on food aid and trade with a diminishing number of friendly countries.

North Korea experienced a horrific period of mass starvation from 1994 to 1998. International isolation, the loss of food aid, cheap oil from the Soviet Union, devastating floods, droughts, and economic mismanagement resulted in a famine that killed as many as a half million people per year. Threats of famine reemerged in 2010, and there are fears that it could easily happen again.

Solid evidence of the current situation is difficult to come by due to the closed economy. Still, South Korean and world health experts believe that starvation deaths still occur in some parts of the country. Much of the population was undernourished even before the COVID-19 pandemic, and three more years of closed borders and isolation can only have made things worse.

Kim Jong-un recently called for a "fundamental transformation" in farming and state economic plans, doubling down on communism's failed theory of centralized control. His rotund figure does not seem to be sufficient cause to question his domination of a malnourished people.

64

ALBANIA—
THE HERMIT KINGDOM

Hermits have no peer pressure.
—STEVEN WRIGHT

ALBANIA IS LOCATED on the Adriatic Sea in southeastern Europe. Occupied by Italy before World War II and Nazi Germany during the war, it subsequently became a one-party communist state (the People's Socialist Republic of Albania), dominated by dictator Enver Hoxha for four decades until he died in 1985.

Albania became known as the "Hermit Kingdom" under Hoxha's rule. Hoxha's delusions and paranoia resulted in ruthless control of a society that saw threats everywhere. Hoxha was deeply afraid of invasions by the Soviet Union, NATO, the United States, Yugoslavia,

Greece, and others. For many years, no one was allowed to leave the country, and many risked their lives to escape.

Albania became a police state, where you spied on your friends, family, and neighbors or were spied upon (or both). Hoxha waged a decades-long campaign of repression against Christian and Muslim believers. Clerics were jailed and even executed. Hoxha proclaimed Albania the world's first atheist state in 1967. The secular realm was not much better, as Hoxha routinely and ruthlessly purged and killed many of his colleagues whom he feared might threaten his hold on the reins of power.

The most striking symbol of Hoxha's paranoia still exists today. The regime created a network of over 750,000 concrete and steel defensive bunkers (in a country roughly the size of Maryland). That is nearly 15 bunkers for every square mile, not just on the borders but throughout the nation. This project was a massive drain on the Albanian economy.

In the event of an outside attack, Albanian reservists were expected to take to the bunkers and defend their homeland. Their training was poor, fuel was scarce, equipment was poor quality, and the government issued antiquated rifles with no ammunition.

It would require extraordinary delusion to conclude that bunkers full of unarmed civilians without means of communication or resupply would be useful in defense. Many of these bunkers have since been torn down and sold for scrap, but thousands still exist nationwide. Some larger bunkers have been repurposed as shelters, storehouses, or even restaurants and cafés.

The communist regime in Albania started to collapse in the early 1990s, leading to economic and social unrest. This resulted in an Albanian diaspora estimated at more than 800,000 people (out of a population of 2.8 million). Albanians joke that there are more Albanians outside the country than inside.

Albania is just beginning to recover from fifty years of isolation and repression. Today, it is a parliamentary constitutional republic rich in natural resources. In 2009, it gained full membership in NATO and has applied to join the European Union. Freed of Hoxa's delusion, Albania can look forward to significant economic growth and enhanced access to regional and global markets.

65

THE COMMUNIST DELUSION

All animals are equal, but some are more equal than others.
—**GEORGE ORWELL**, Animal Farm

COMMUNISM is a political and economic ideology that envisions a working-class revolution leading to a classless society where the state (rather than individuals) owns and controls property and economic resources. People often lump socialism and communism together, but while similar, these political ideologies are distinct in two ways. First, communism sees violent worker revolution against the upper classes as an essential element for achieving a communist state. Socialism is a little more flexible. Second, socialism allows for the ownership of private property by the individual, with the means of production and distribution of resources controlled by a democratically elected government.

The Communist Manifesto, published in 1848, outlines Karl Marx and Friedrich Engels's theories of society, politics, class struggle, and the failures of capitalism and socialism. Communism emerged partly in response to the cruel exploitation of workers and the yawning gap between rich and poor that were byproducts of the Industrial Revolution.

Communism appealed to fundamental notions of fairness. From each according to their abilities and to each according to their needs. These ideals resonate with compassionate people. It feels good to care about the least fortunate and be fair and equal. The dog-eat-dog ethic of capitalism seems cruel by comparison.

However, three significant problems with communism demonstrate the delusion of viewing it as a viable economic and social alternative. The first is the idea that economic unfairness justifies revolution, murder, theft, and loss of personal freedom. In other words, the goal of a classless society is morally important enough that any method of achieving it is acceptable. The ends justify the means because we care (as long as we remain in charge and on the right end of the rifle).

In the Soviet Union, tens of millions were killed by Stalin's effort to build a utopian communist society. Mao Zedong killed millions more in China to secure power and impose his vision of communism. Ho Chi Minh sacrificed hundreds of thousands (some estimates say millions) to achieve his dream of a communist Vietnam. Fidel Castro and Che Guevara built and exported their Marxist revolution on the recruiting technique of join us, or we kill you.

The second major problem with communism is its inevitable tilt towards dictatorship and totalitarianism. When one party controls every detail of the economy and society, opposition cannot be tolerated. Opponents are ruthlessly suppressed. The press becomes an organ of the state. Secret police are given free rein to do whatever is necessary to maintain control.

Even communist societies that accept socialist notions of private property (like Vietnam and China) have one rule that stands above all others. The sin that cannot be forgiven is challenging the ruling party's power and control. Any threat to the ruling class will be crushed with overwhelming force. Dissidents are jailed, and their families are punished, sometimes even to the next generation.

A corollary to this is that communist leaders tend to live lavish lifestyles, with palaces, summer homes, limousines, and mistresses. Vladimir Putin has an estimated net worth of $70 billion. Most billionaires in China are Communist Party members or officials. All animals are equal, but some are more equal than others.

The third major problem with communism is that government ownership and control of private property and the means of production inevitably results in lower levels of innovation and productivity. Communism fails to appreciate self-interest as an inherent element of human nature. Communism supposes that results (distribution of resources) should be equal irrespective of effort, ability, opportunity, or even luck.

But fairness relates to effort and ability as much as to outcomes. People quickly realize that if they get the same result no matter how hard they work, they don't feel the need to work very hard. Slackers will slack. It may be unfair if your neighbor has more resources than you. It is also unfair if you work harder than your neighbor and get paid the same. This dynamic inevitably results in productivity declines. Everyone may be equal, but everyone also becomes poor.

Some argue that communism has never really been implemented in its proper form. That may be true. But in every instance where it has been attempted, totalitarianism, suppression, economic decline, and loss of political and personal freedom have been the result.

In free market economies, innovation and choices emerge. People acting in self-interest can see the rewards for their efforts and engage in mutually beneficial transactions. Free markets improve the standard of living by rewarding individuals for innovation and their efforts.

However, free market economics have never really been implemented in their purest form either. The dog-eat-dog excesses of pure free-market capitalism need to be tamed, and this is accomplished in the United States by blending in some of the ideals of social democracy.

By way of example, a free market entrepreneur in the United States recently woke up in his peaceful village that was kept safe from foreign and domestic threats using shared government resources, took a shower using water from a system built by shared government resources, ate breakfast that was kept safe from pathogens by shared government resources, took his medications that were vetted for safety and efficacy by shared government resources, drove to work on roads built and maintained by shared government resources, stopped at a coffee shop to buy coffee using money that was reliable and protected from inflation by shared government resources, entered his business which operated on a level playing field using a system of laws and courts paid for by shared government resources, turned on his computer that connected to the internet that was created and built using shared government resources, and logged on to his social media account to tell the story about how rugged individualism built his business.

In fairness to the entrepreneur, he very likely paid his share of taxes along the way to support all of those local, state, and federal systems. But his business is not operating in a purely free market system. Private ownership is part of it. Free markets are part of it. Enlightened self-interest, initiative, and effort are an essential part of it. But shared, communal resources (central to the concept of social democracy) are also critical to his success.

Democracy and capitalism have been called inadequate systems that are better than any other. At a minimum, they are less delusional.

EPILOGUE

No man is happy without a delusion of some kind.
—CHRISTIAN NESTELL BOVEE

YOUR LIFE

Thoughtful alternative medicine practitioners may one day prove me wrong in my broadly negative assessment of their practices. We may discover precisely how acupuncture works so that we can use it with precision to treat illness reliably. Large-scale clinical trials may one day help us understand how spinal manipulation techniques can help cure migraine headaches or ear infections.

But even if this never happens, alternative medicine will continue to be practiced despite my low opinion. That may be okay. If alternative medicine produces a placebo effect and the patient feels better, that is a good thing. Trying something new or experimental is not unreasonable when traditional medicine fails and you are suffering. Alternative medicine can provide hope and serve as an alternative to despair. You be the judge.

YOUR MONEY

When the next sequel to Extraordinary Popular Delusions is written 200 years from now, the author may include chapters on how the lack of critical thinking or evidence-based science led to economic or ecological collapse.

Investing for retirement using U.S. currency today seems prudent and proper, yet it might one day be considered delusional. Heating our homes with natural gas is common today, but we may be deluding ourselves about the environmental costs. Venting on social media about the complete insanity of people who disagree with our political opinions is wildly popular, yet it is perhaps the most delusional thinking of all.

Yet, with all that delusion, we continue to lead our lives and are generally content with our situation. There are risks in how we save for retirement, and there are risks in not saving for retirement. It is not always clear what the best options are, so you make your choices and you take your chances.

YOUR SOUL

Faith can be a force for good as well as a force for evil. On balance (with some obvious exceptions), it is a force for good in the world. Some believe that moral and ethical behavior is impossible (or at least unlikely) without a foundation of faith. I come down on the other side of this argument. Agnostics Bertrand Russell, Winston Churchill, Paul McCartney, and Albert Einstein seem like decent human beings. Osama bin Laden was both pious and evil.

I have never met anyone with precisely the same beliefs as I have. Occasionally, people with different opinions help me see things in a new light and change my views. While there is a chance that everyone else is wrong and my beliefs are perfect, the accumulated evidence tells me those are very long odds.

In reflecting on your beliefs (and your delusions), may the odds be in your favor.

NOTES AND SOURCES

VOLUME 1—YOUR LIFE (HEALTH AND WELLNESS DELUSIONS)

THE ANTI-VACCINE MOVEMENT

World Health Organization, "Top Ten Threats to Global Health in 2019"
  http://www.who.int/emergencies/ten-threats-to-global-health-in-2019

History of Vaccines, "History of Anti-Vaccination Movements"
  https://historyofvaccines.org/vaccines-101/misconceptions-about-vaccines/history-anti-vaccination-movements

World Health Organization, "Vaccines and Immunization," October 29, 2019
  https://www.who.int/health-topics/vaccines-and-immunization#tab=tab_1

Reich, Jennifer, "Measles 2019: Modern Parenting Culture Fuels Anti-Vaccination Movement," Vox, June 13, 2019
  https://www.vox.com/first-person/2019/5/8/18535944/jessica-biel-measles-2019-outbreak-anti-vax

Baker, Jeffrey P, "Mercury, Vaccines, and Autism," American Journal of Public Health, American Public Health Association, February 1, 2008
  https://doi.org/10.2105/ajph.2007.113159

Deer, Brian, "How the Case against the MMR Vaccine Was Fixed." BMJ, January 5, 2011
  https://doi.org/10.1136/bmj.c5347

Bergengruen, Vera, "How the Anti-Vax Movement Is Taking Over the Right," Time, January 26, 2022
  https://time.com/6141699/anti-vaccine-mandate-movement-rally/

Crislip, Mark, "Homeopathic Vaccines" Science-Based Medicine, November 6, 2010
  https://sciencebasedmedicine.org/homeopathic-vaccines/

International Chiropractors Association, "ICA Affirms Policy on Health Freedom," October 12, 2021
  https://www.chiropractic.org/ica-affirms-policy-on-health-freedom/

Centers for Disease Control and Prevention, "Global Polio Eradication," September 27, 2023
  https://www.cdc.gov/polio/global-polio-eradication.html

NOTES AND SOURCES

"GPEI-Endemic Countries"
https://polioeradication.org/where-we-work/polio-endemic-countries/

ALTERNATIVE MEDICINE

Johns Hopkins Medicine, "Types of Complementary and Alternative Medicine," November 19, 2019
https://www.hopkinsmedicine.org/health/wellness-and-prevention/types-of-complementary-and-alternative-medicine

Angell, Marcia, and Kassirer, Jerome P., "Alternative Medicine—The Risks of Untested and Unregulated Remedies," The New England Journal of Medicine. Massachusetts Medical Society, September 17, 1998
https://doi.org/10.1056/nejm199809173391210

Gorski, David, "'Integrative Medicine': A Brand, Not a Specialty," Science-Based Medicine, September 20, 2017
https://sciencebasedmedicine.org/integrative-medicine-a-brand-not-a-specialty/

Ernst, Edzard, "How Much of CAM Is Based on Research Evidence?," Hindawi Publishing Corporation, January 1, 2011
https://doi.org/10.1093/ecam/nep044

"$2.5 Billion Spent, No Alternative Cures Found," June 10, 2009
https://www.nbcnews.com/id/wbna31190909

"Psychology and 'Alternative Medicine'"
https://web.archive.org/web/20111012091930/http://www.sram.org/0302/bias.html

Beyerstein, Barry L. PhD, "Alternative Medicine and Common Errors of Reasoning : Academic Medicine," March 2001,
https://journals.lww.com/academicmedicine/fulltext/2001/03000/alternative_medicine_and_common_errors_of.9.aspx

Ernst, Edzard, "Alternative Medicine Remains an Ethics-Free Zone." The Guardian, September 20, 2017
https://www.theguardian.com/science/blog/2011/nov/08/alternative-medicine-ethics-free-zone

CHIROPRACTIC

Kirkey, Sharon, and Brice Hall, "The First Chiropractor Was a Canadian Who Claimed He Received a Message from a Ghost." thestarphoenix, July 2, 2019
https://thestarphoenix.com/health/the-first-chiropractor-was-a-canadian-who-claimed-he-received-a-message-from-a-ghost/wcm/e419ac76-d4d4-4f19-b2b3-448ea647eb57

"A Close Look at Chiropractic Wrongdoing | Quackwatch," August 24, 2022
https://quackwatch.org/chiropractic/general/

Côté, Pierre, André Bussières, J. David Cassidy, Jan Hartvigsen, Greg Kawchuk, Charlotte Leboeuf–Yde, Silvano Mior, and Michael J. Schneider. "A United Statement of the Global Chiropractic Research Community against the Pseudoscientific Claim That Chiropractic Care Boosts Immunity." Chiropractic & Manual Therapies. BioMed Central, May 4, 2020
https://doi.org/10.1186/s12998-020-00312-x

clinicaltrials.gov, "CTG Labs—NCBI"
https://clinicaltrials.gov/search?intr=chiropractic

Homola, D.C., Samuel, Inside Chiropractic: A Patient's Guide. Prometheus Books, 1999 ISBN 1573926981

Quackwatch.org, "Undercover Investigations of Chiropractors" March 21, 2004
https://quackwatch.org/chiropractic/research/chiroinv/

LeFebvre, Ron, and David H. Peterson and Mitchell Haas, "Evidence-Based Practice and Chiropractic Care." Journal of Evidence-Based Complementary & Alternative Medicine. SAGE Publishing, September 3, 2012
https://www.ncbi.nlm.nih.gov/pmc/articles/PMC3716373/

HOMEOPATHY

Mukerji, Nikil, and Edzard Ernst, "Why Homoeopathy Is Pseudoscience." Springer Science+Business Media, September 14, 2022
https://doi.org/10.1007/s11229-022-03882-w

"Helios Homoeopathy"
https://www.helios.co.uk/shop/search/remedies

European Academies' Science Advisory Council, "Homeopathic Products and Practices: Assessing the Evidence and Ensuring Consistency in Regulating Medical Claims in the EU," September 2017
https://easac.eu/fileadmin/PDF_s/reports_statements/EASAC_Homepathy_statement_web_final.pdf

Loudon, Irvine, "A Brief History of Homeopathy," Journal of the Royal Society of Medicine, SAGE Publishing, December 1, 2006
https://doi.org/10.1258/jrsm.99.12.607

Center ForInquiry, "Consumers Feel 'Scammed' by Walmart and CVS over Homeopathic Fake Medicine, Survey Shows," June 7, 2022
https://centerforinquiry.org/press_releases/consumers-feel-scammed-by-walmart-and-cvs-over-homeopathic-fake-medicine/

Barrett, M.D., Stephen, "Homeopathy: The Ultimate Fake | Quackwatch," August 25, 2016
https://quackwatch.org/related/homeo/

Kolton, Eleanor A, "FDA Toughens Enforcement of Homeopathic Products," www.natlawreview.com, August 25, 2023
https://www.natlawreview.com/article/fda-toughens-enforcement-homeopathic-products

NATUROPATHY

Encyclopedia.Com, "Naturopathic Medicine"
https://www.encyclopedia.com/medicine/divisions-diagnostics-and-procedures/medicine/naturopathic-medicine

Barrett, Stephen, Quackwatch.org, "A Close Look at Naturopathy," November 26, 2013
https://quackwatch.org/related/naturopathy/

NOTES AND SOURCES

Barrett, Stephen, Quackwatch.org, "Naturopathic Opposition to Immunization," May 31, 2019
https://quackwatch.org/related/Naturopathy/immu/

Chivers, Tom, Spectator Health, "How Does Naturopathy Work? A Bit like a Flying Vacuum-Cleaner to Mars," June 16, 2015
https://web.archive.org/web/20170905094710/https://health.spectator.co.uk/how-does-naturopathy-work-a-bit-like-a-flying-vacuum-cleaner-to-mars/

Devlin, Hannah, "The Naturopath Whistleblower: 'It Is Surprisingly Easy to Sell Snake Oil,'" The Guardian, April 10, 2018
https://www.theguardian.com/lifeandstyle/2018/mar/27/naturopath-whistleblower-snake-oil-multi-billion-dollar

Gorski, David, Science-Based Medicine, "Naturopathy and Science" December 19, 2016
https://sciencebasedmedicine.org/naturopathy-and-science/

ACUPUNCTURE

Li, Zhisui, The Private Life of Chairman Mao, New York: Random House, 1996

Pyne, D., and N.G. Shenker. "Demystifying Acupuncture," May 6, 2008
https://academic.oup.com/rheumatology/article/47/8/1132/1786327?login=false

Quackwatch.org, "Acupuncture," August 14, 2019
https://quackwatch.org/acupuncture/

"Be Wary of Acupuncture, Qigong, and 'Chinese Medicine,'" December 6, 2022
https://quackwatch.org/related/acu/

Ingraham, Paul, "Acupuncture as a Pain Treatment: A Skeptical Perspective," www.PainScience.com, January 15, 2022
https://www.painscience.com/articles/acupuncture-for-pain.php

Hutchinson, Amanda J P, Simon Ball, Jeremy C H Andrews, and Gareth Jones. "The Effectiveness of Acupuncture in Treating Chronic Non-Specific Low Back Pain: A Systematic Review of the Literature," Journal of Orthopaedic Surgery and Research, BioMed Central, January 1, 2012
https://doi.org/10.1186/1749-799x-7-36

Colquhoun, PhD, David, and Steven P. Novella, MD. "Acupuncture Is Theatrical Placebo," 2013
http://www.dcscience.net/Colquhoun-Novella-A%26A-2013.pdf

Crislip, Mark, Science-Based Medicine, "Acupuncture Odds and Ends," January 12, 2017
https://sciencebasedmedicine.org/acupuncture-odds-and-ends/.

ESSENTIAL OILS

"Dos and Don'ts of Essential Oils," WebMD
https://www.webmd.com/skin-problems-and-treatments/ss/slideshow-essential-oils

"Essential Oil Profile Directory—Uses and Benefits for Over 130 Oils," AromaWeb," https://www.aromaweb.com/essentialoils/index.php

NOTES AND SOURCES

Schneider, Kim, "11 Essential Oils: Their Benefits and How To Use Them," Cleveland Clinic, July 27, 2023
https://health.clevelandclinic.org/essential-oils-101-do-they-work-how-do-you-use-them/

U.S. Food and Drug Administration, "Development & Approval Process | Drugs." U.S. Food and Drug Administration, August 8, 2022
https://www.fda.gov/drugs/development-approval-process-drugs

U.S. Food and Drug Administration, "Products and Medical Procedures," September 14, 2021
https://www.fda.gov/medical-devices/products-and-medical-procedures

"Aromatherapy: Do Essential Oils Really Work?," Johns Hopkins Medicine, August 8, 2021
https://www.hopkinsmedicine.org/health/wellness-and-prevention/aromatherapy-do-essential-oils-really-work

"Essential Oils: Poisonous When Misused," Poison Control
https://www.poison.org/articles/essential-oils.

U.S. Food and Drug Administration, Center For Food Safety And Applied Nutrition, "Aromatherapy" September 8, 2023
https://www.fda.gov/cosmetics/cosmetic-products/aromatherapy

Healthline, "Tea Tree Oil for Eczema Flare-Ups: Benefits, Risks, and More," July 10, 2023
https://www.healthline.com/health/skin-disorders/tea-tree-oil-for-eczema#takeaway

U.S. Food and Drug Administration, Office Of Regulatory Affairs, "Advisory Letters," June 16, 2023
https://www.fda.gov/inspections-compliance-enforcement-and-criminal-investigations/compliance-actions-and-activities/advisory-letters

U.S. Food and Drug Administration, Center For Food Safety And Applied Nutrition, "Young Living Essential Oils Corporate—615777—06/10/2022," June 10, 2022
https://www.fda.gov/inspections-compliance-enforcement-and-criminal-investigations/warning-letters/young-living-essential-oils-corporate-615777-06102022

DIETARY SUPPLEMENTS

National Institutes of Health, Office of Dietary Supplements, "Dietary Supplements: What You Need to Know"
https://ods.od.nih.gov/factsheets/WYNTK-Consumer/

U.S. Food and Drug Administration, Center For Food Safety And Applied Nutrition, "Dietary Supplements" March 6, 2023
https://www.fda.gov/food/dietary-supplements

Federal Trade Commission, "Three Dietary Supplement Marketers Settle FTC, Maine AG Charges," September 18, 2021
https://www.ftc.gov/news-events/news/press-releases/2017/08/three-dietary-supplement-marketers-settle-ftc-maine-ag-charges

"Dietary Supplements Market Size, Share & Trends Analysis Report By Ingredient, By Type, By End-User, By Distribution Channel, By Form, By Application, By Region, And Segment Forecasts, 2023—2030"
https://www.grandviewresearch.com/industry-analysis/dietary-supplements-market

Navarro, Victor J., Ikhlas A. Khan, Einar Bjornsson, Leonard B. Seeff, Jose Serrano, and Jay H. Hoofnagle. "Liver Injury from Herbal and Dietary Supplements." Hepatology. Wiley, November 17, 2016
https://doi.org/10.1002/hep.28813

Quackwatch.org, "'Dietary Supplements,' Herbs, and Hormones," February 28, 2005
https://quackwatch.org/related/suppsherbs/

Quackwatch.org, "The Herbal Minefield," November 23, 2013
https://quackwatch.org/related/herbs/

Harmon, Katherine, "Herbal Supplement Sellers Dispense Dangerous Advice, False Claims," Scientific American, May 28, 2010
https://www.scientificamerican.com/article/herbal-supplement-dangers/

U.S. Food and Drug Administration, Office Of Regulatory Affairs, "Health Fraud Product Database"
https://www.fda.gov/consumers/health-fraud-scams/health-fraud-product-database

DETOX SCHEMES

Harvard Health, "The Dubious Practice of Detox," May 1, 2008
https://www.health.harvard.edu/staying-healthy/the-dubious-practice-of-detox

Aronsohn, Andrew "Is Detoxing Good for You?" UChicago Medicine, December 21, 2022
https://www.uchicagomedicine.org/forefront/gastrointestinal-articles/do-detoxes-work

Van De Walle, Gavin, "What Is a Full-Body Detox?" Healthline, May 16, 2023
https://www.healthline.com/nutrition/how-to-detox-your-body

Mayo Clinic, "Colon Cleansing: Is It Helpful or Harmful?," May 24, 2022
https://www.mayoclinic.org/healthy-lifestyle/consumer-health/expert-answers/colon-cleansing/faq-20058435

Barrett, Stephen, "The Detox Foot Pad Scam" Quackwatch, November 12, 2010
https://quackwatch.org/device/reports/kinoki/

American Council on Science and Health, "What You Don't Know About Formaldehyde Will Leave You Floored," February 29, 2016
https://www.acsh.org/news/2016/02/25/stuff-you-didnt-know-about-formaldehyde

Digestive HealthTeam, "Do Detoxes and Cleanses Actually Work?" Cleveland Clinic, March 22, 2023
https://health.clevelandclinic.org/are-you-planning-a-cleanse-or-detox-read-this-first/

Robinson, Kara Mayer, "The Lemonade Diet/Master Cleanse," WebMD, December 12, 2013
https://www.webmd.com/diet/a-z/lemonade-master-cleanse-diet

U.S Food and Drug Administration, "FDA Warns Marketers of Unapproved 'Chelation' Drugs"
https://wayback.archive-it.org/7993/20170111123610/http://www.fda.gov/ForConsumers/ConsumerUpdates/ucm229358.htm

Gorski, David, "The Result of the Trial to Assess Chelation Therapy (TACT): As Underwhelming as Expected" Science-Based Medicine, January 12, 2017
https://sciencebasedmedicine.org/the-result-of-the-trial-to-assess-chelation-therapy-tact-as-underwhelming-as-expected/

Petrie, Tim, "Do Detox Foot Pads Really Work?" Verywell Health, February 8, 2023
https://www.verywellhealth.com/detox-foot-pads-5221565

Leviatan, Noam, and Ofir Kuperman, Davidson Institute of Science Education, "A Lie Has Legs—Detox Foot Pads," September 1, 2022
https://davidson.weizmann.ac.il/en/online/reasonabledoubt/lie-has-legs

Ernst, E. M.D., Ph.d., "Colonic Irrigation and the Theory of Autointoxication: A Triumph of Ignorance Over Science," Journal of Clinical Gastroenterology," June, 1997
https://journals.lww.com/jcge/fulltext/1997/06000/colonic_irrigation_and_the_theory_of.2.aspx

CUPPING

DC's Improbable Science, "Cupping: Bruises for the Gullible, and Other Myths in Sport," August 12, 2016
http://www.dcscience.net/2016/08/10/cupping-bruises-for-the-gullible-and-other-myths-in-sport/

Novella, Steven, Science-Based Medicine, "Cupping—Olympic Pseudoscience," November 25, 2017
https://sciencebasedmedicine.org/cupping-olympic-pseudoscience/

Shmerling, Robert H, Harvard Health, "What Exactly Is Cupping?" June 22, 2020
https://www.health.harvard.edu/blog/what-exactly-is-cupping-2016093010402

EAR CANDLING

Dancer, Heather L. and Jess Shenk, "Ear Candling: A Fool Proof Method, or Proof of Foolish Methods?," AudiologyOnline
https://www.audiologyonline.com/articles/ear-candling-fool-proof-method-1010

Roazen, Lisa, Quackwatch.org, "Why Ear Candling Is Not a Good Idea," May 12, 2010
https://quackwatch.org/related/candling/

U.S. Food and Drug Administration, "Import Alert 77-01,"
https://www.accessdata.fda.gov/cms_ia/importalert_225.html

REIKI

Bellamy, Jann, Science-Based Medicine, "Reiki: Fraudulent Misrepresentation," January 3, 2017
https://sciencebasedmedicine.org/reiki-fraudulent-misrepresentation/

Dunning, Brian, Skeptoid, "Your Body's Alleged Energy Fields"
https://skeptoid.com/episodes/4411

Barrett, Stephen, Quackwatch, "Reiki Is Nonsense," March 3, 2022
https://quackwatch.org/related/reiki/

Rosa, Linda. "A Close Look at Therapeutic Touch," JAMA, American Medical Association, April 1, 1998
https://doi.org/10.1001/jama.279.13.1005

Barrett, Stephen, Quackwatch, "Why Therapeutic Touch Should Be Considered Quackery," February 3, 2008
https://quackwatch.org/related/tt/

Bechtel, William, and Robert C. Richardson, "Vitalism," 1998, Routledge Encyclopedia of Philosophy
http://mechanism.ucsd.edu/teaching/philbio/vitalism.htm

ALTERNATIVE MEDICINE FOR PETS

Ramey, David, Science-Based Medicine, "Animal Acupuncture," June 8, 2009
https://sciencebasedmedicine.org/animal-acupuncture/

Gorski, David , "'Cat-Upuncture'? What Did Those Poor Cats Ever Do to Deserve This?" ScienceBlogs, March 4, 2016
https://scienceblogs.com/insolence/2016/03/04/cat-upuncture-what-did-those-poor-cats-ever-do-to-deserve-this

Lees, P., Ludovic Pelligand, Martin Whiting, Duncan Chambers, P. L. Toutain, and Whitehead M.L., "Comparison of Veterinary Drugs and Veterinary Homeopathy: Part 2." Veterinary Record. John Wiley & Sons Ltd, August 18, 2017
https://doi.org/10.1136/vr.104279

Ebani, Valentina Virginia, and Francesca Mancianti, "Use of Essential Oils in Veterinary Medicine to Combat Bacterial and Fungal Infections," Veterinary Sciences, Multidisciplinary Digital Publishing Institute, November 30, 2020
https://doi.org/10.3390/vetsci7040193

Barrett, Stephen, Quackwatch.org, "Does Homeopathy Work in Animals?," January 5, 2018
https://quackwatch.org/homeopathy/articles/animals/

Barrett, Stephen, Quackwatch.org, "Veterinary Chiropractic," December 15, 2000
https://quackwatch.org/chiropractic/dd/chirovet/

FAD DIETS

"Fad Diets." British Dietetic Association
https://www.bda.uk.com/resource/fad-diets.html

Federal Trade Commission, "The Truth Behind Weight Loss Ads," August 17, 2022,
https://consumer.ftc.gov/articles/truth-behind-weight-loss-ads

Joshi, Suvarna, and Viswanathan Mohan, "Pros & Cons of Some Popular Extreme Weight-Loss Diets," Indian Journal of Medical Research. Medknow, January 1, 2018
https://doi.org/10.4103/ijmr.ijmr_1793_18

Wdowik, Melissa, "The Long, Strange History of Dieting Fads," Colorado State University, The Conversation, November 13, 2017
https://source.colostate.edu/the-long-strange-history-of-dieting-fads/

"Popular Diet Trends: Today's Fad Diets" Today's Dietitian Magazine,
https://www.todaysdietitian.com/newarchives/0519p12.shtml

Lynch, Rene, "A Brief Timeline Shows How We're Gluttons for Diet Fads" Los Angeles Times, Los Angeles Times, February 28, 2015
https://www.latimes.com/health/la-he-diet-timeline-20150228-story.html

NOTES AND SOURCES

Quackwatch.org, "Impossible Weight-Loss Claims Summary of an FTC Report," December 16, 2003
https://quackwatch.org/related/PhonyAds/weightlossfraud/

Horton, John, "Does the Grapefruit Diet Work?" Cleveland Clinic, July 27, 2021
https://health.clevelandclinic.org/grapefruit-diet/

Sheehan, Stephen, "The Ultimate Guide to the Bulletproof Diet," Bulletproof, February 14, 2022
https://www.bulletproof.com/diet-articles/ultimate-bulletproof-diet-guide/

Cusack, Leila, Emmy De Buck, Veerle Compernolle, and Philippe Vandekerckhove, "Blood Type Diets Lack Supporting Evidence: A Systematic Review." The American Journal of Clinical Nutrition, Elsevier BV, July 1, 2013
https://doi.org/10.3945/ajcn.113.058693

BIODYNAMICS

Keyser, Raini, "Biodynamic Farming and Wine," Vinum 55, October 3, 2022
https://www.vinum55.com/biodynamic-farming-wine/

Chhabra, Esha, "Biodynamic Farming Is on the Rise—but How Effective Is This Alternative Agricultural Practice?" the Guardian, August 2, 2018
https://www.theguardian.com/sustainable-business/2017/mar/05/biodynamic-farming-agriculture-organic-food-production-environment

Novella, Steven, "Biodynamic Farming and Other Nonsense," NeuroLogica Blog—Your Daily Fix of Neuroscience, Skepticism, and Critical Thinking, June 19, 2017
https://theness.com/neurologicablog/index.php/biodynamic-farming-and-other-nonsense/

"Biodynamic Association | Rethinking Agriculture"
https://www.biodynamics.com/

Capretti, Lucia, "Understanding the Lunar Calendar in Biodynamic Viticulture—SOMM TV Magazine." SOMM TV Magazine—A World of Wine, Food & Travel, January 16, 2023
https://mag.sommtv.com/2023/01/lunar-calendar-biodynamic-viticulture/

MONSANTO AND THE GMO DELUSION

Senapathy, Kavin, "I Was Lured Into Monsanto's GMO Crusade. Here's What I Learned." Undark Magazine, September 30, 2019
https://undark.org/2019/06/27/monsanto-gmo-crusade/

Wellness Resources,
https://www.wellnessresources.com/news/health-scandal-of-the-decade-monsantos-gmo-perversion-of-food

Kush, Stephanie, "Why I No Longer Believe That Monsanto Is The Devil." Illinois Farm Families, June 8, 2016,
https://watchusgrow.org/2016/06/08/why-i-no-longer-believe-that-monsanto-is-the-devil/

Bawa, A. S., and K.R. Anilakumar, "Genetically Modified Foods: Safety, Risks and Public Concerns—a Review," Journal of Food Science and Technology, Springer Science+Business Media, December 19, 2012
https://doi.org/10.1007/s13197-012-0899-1

Burger, Ludwig, "With Deal to Close This Week, Bayer to Retire Monsanto Name" June 4, 2018, https://www.reuters.com/article/us-monsanto-m-a-bayer-closing/with-deal-to-close-this-week-bayer-to-retire-monsanto-name-idUSKCN1J00IZ

Rosenberg, Meredith, "GMOs: Everything You Need to Know." EcoWatch, November 26, 2022 https://www.ecowatch.com/understanding-gmos-2653417556.html

VOLUME 2—YOUR MONEY (FINANCIAL DELUSIONS)

DOT-COM BUBBLE

"Jobfairy.Com—Boo! And the 100 Other Dumbest Moments in e-Business History," https://www.jobfairy.com/articles01/BooAndthe100OtherDumbestM.html

Lanxon, Nate, "The Greatest Defunct Web Sites and Dotcom Disasters." CNET, November 18, 2009 https://www.cnet.com/tech/computing/the-greatest-defunct-web-sites-and-dotcom-disasters/

"Here's Why The Dot Com Bubble Began And Why It Popped." Business Insider, December 16, 2010 https://www.businessinsider.com/heres-why-the-dot-com-bubble-began-and-why-it-popped-2010-12

Dai, Shackelford and Zhang, "Capital Gains Taxes and Stock Return Volatility: Evidence from the Taxpayer Relief Act of 1997" https://www.businessinsider.com/heres-why-the-dot-com-bubble-began-and-why-it-popped-2010-12

Beer, Jeff, "20 Years Ago, the Dot-Coms Took over the Super Bowl." Fast Company, January 21, 2020 https://www.fastcompany.com/90453258/20-years-ago-the-dot-coms-took-over-the-super-bowl

"AOL and Time Warner to Merge—Jan. 10, 2000," January 10, 2000 https://money.cnn.com/2000/01/10/deals/aol_warner/

"Nasdaq, Dow Take Nosedive—Apr. 14, 2000," April 14, 2000 https://money.cnn.com/2000/04/14/markets/markets_newyork/

"Dot.Coms Lose $1.755 Trillion in Market Value—Nov. 9, 2000," November 9, 2000 https://money.cnn.com/2000/11/09/technology/overview/

Beltran, Luisa, CNN/Money, "WorldCom Files Largest Bankruptcy Ever—Jul. 19, 2002," July 19, 2002 https://money.cnn.com/2002/07/19/news/worldcom_bankruptcy/

"Webvan Announces Shutdown, Chapter 11 Filing—Jul. 9, 2001," July 9, 2001 https://money.cnn.com/2001/07/09/technology/webvan/

"10 Big Dot.Com Flops—Pets.Com (1)—CNNMoney.Com," n.d. https://money.cnn.com/galleries/2010/technology/1003/gallery.dot_com_busts/

"10 Big Dot.Com Flops—eF.Com (3)—CNNMoney.Com," n.d. https://money.cnn.com/galleries/2010/technology/1003/gallery.dot_com_busts/3.html

"PETA v. Doughney," n.d. https://cyber.harvard.edu/stjohns/PETA_v_Doughney.html

"Remembering Netscape: The Birth Of The Web—July 25, 2005," July 25, 2005 https://web.archive.org/web/20060427112146/http://money.cnn.com/magazines/fortune/fortune_archive/2005/07/25/8266639/index.htm

MacroTrends. "NASDAQ Composite—45 Year Historical Chart," n.d.
https://www.macrotrends.net/1320/nasdaq-historical-chart

Justia Law, "United States v. Microsoft Corp., 87 F. Supp. 2d 30 (D.D.C. 2000)," n.d.
https://law.justia.com/cases/federal/district-courts/FSupp2/87/30/2307082/

Brookings, "The Telecommunications Crash: What To Do Now? | Brookings," July 28, 2016
https://www.brookings.edu/articles/the-telecommunications-crash-what-to-do-now/

Bartash, Jeffry, "Anatomy of Global Crossing's Failure." MarketWatch, January 29, 2002
https://www.marketwatch.com/story/anatomy-of-global-crossings-failure

Roberts, Joel, "Time Warner Drops AOL Name." CBS News, October 22, 2003
https://www.cbsnews.com/news/time-warner-drops-aol-name/.

Davis, Andrew, "At One Point, Amazon Lost More than 90% of Its Value. But Long-Term Investors Still Got Rich." CNBC, December 18, 2018
https://www.cnbc.com/2018/12/18/dotcom-bubble-amazon-stock-lost-more-than-90percent-long-term-investors-still-got-rich.html

"The Insane History of Beanie Babies," n.d.
https://www.withotis.com/mag/beanie-babies-boom

The Fiscal Times, "How the Great Beanie Baby Bubble Went Bust," March 2, 2015 https://www.thefiscaltimes.com/2015/03/02/How-Great-Beanie-Baby-Bubble-Went-Bust

US HOUSING BUBBLE AND GLOBAL FINANCIAL CRISIS (2007-08)

Lewis, Michael, The Big Short, 2015. ISBN 978-0-393-07223-5

Duffie, Darrell, "Prone to Fail: The Pre-Crisis Financial System." Journal of Economic Perspectives, American Economic Association, February 1, 2019
https://doi.org/10.1257/jep.33.1.81

NCRC, "Don't Blame the Affordable Housing Goals for the Financial Crisis." February 11, 2019
https://ncrc.org/dont-blame-affordable-housing-goals-financial-crisis/

Maverick, J.B., "Consequences of the Glass-Steagall Act Repeal." Investopedia, September 9, 2023
https://www.investopedia.com/ask/answers/050515/did-repeal-glasssteagall-act-contribute-2008-financial-crisis.asp

Aaron, Kat, "Predatory Lending: A Decade of Warnings." Center for Public Integrity, April 11, 2022
https://publicintegrity.org/inequality-poverty-opportunity/predatory-lending-a-decade-of-warnings/

Amadeo, Kimberly, "2009 Financial Crisis Explanation with Timeline." The Balance, January 30, 2021
https://www.thebalancemoney.com/2009-financial-crisis-bailouts-3305539

Guillén, Mauro, "The Global Economic & Financial Crisis: A Timeline "
https://lauder.wharton.upenn.edu/wp-content/uploads/2015/06/Chronology_Economic_Financial_Crisis.pdf.

David Ellis, "Feds Unveil Rescue Plan for Fannie, Freddie—Sep. 7, 2008," September 7, 2008
https://money.cnn.com/2008/09/07/news/companies/fannie_freddie/

Calabria, Mark, "Fannie, Freddie, and the Subprime Mortgage Market." cato.org, March 7, 2011
https://www.cato.org/sites/cato.org/files/pubs/pdf/bp120.pdf

Adler, Lynn, "Foreclosures Soar 81 Percent in 2008." U.S., January 15, 2009
https://www.reuters.com/article/us-usa-mortgages-foreclosures/foreclosures-soar-81-percent-in-2008-idUSTRE50E1KV20090115

Hansen, Chris and Greenberg, Richard, NBC News, "If You Had a Pulse, We Gave You a Loan," March 22, 2009
https://www.nbcnews.com/id/wbna29827248

Mollenkamp, Carrick, Craig, Suzanne, Ng, Serena, and Lucchetti, Aaron, "Lehman Files for Bankruptcy." online.wsj.com, September 16, 2008
https://web.archive.org/web/20101204201026/http://online.wsj.com/article/SB122145492097035549.html

GHOST CITIES—THE CHINESE HOUSING BUBBLE

"Concrete 'Ghost Towns' Make China's Real Estate Bubble Visible," Nikkei Asia, February 9, 2022
https://asia.nikkei.com/Spotlight/The-age-of-Great-China/Concrete-ghost-towns-make-China-s-real-estate-bubble-visible

Batarags, Lina, "China Has at Least 65 Million Empty Homes—Enough to House the Population of France. It Offers a Glimpse into the Country's Massive Housing-Market Problem." Business Insider, October 14, 2021
https://www.businessinsider.com/china-empty-homes-real-estate-evergrande-housing-market-problem-2021-10

Toh, Michelle, "Ghost Towns: Evergrande Crisis Shines a Light on China's Millions of Empty Homes." October 15, 2021
https://www.cnn.com/2021/10/14/business/evergrande-china-property-ghost-towns-intl-hnk/index.html

Brahambhatt, Rupendra, "How China's Ghost Cities Are Linked to the Evergrande Crisis?," October 29, 2021
https://interestingengineering.com/culture/chinas-ghost-cities-and-its-65-million-empty-homes

Zhou, Christina, Shelton, Tracy and Pan, Ning, "China's Eerie Ghost Cities a 'symptom' of the Country's Economic Troubles and Housing Bubble." ABC News, June 26, 2018
https://www.abc.net.au/news/2018-06-27/china-ghost-cities-show-growth-driven-by-debt/9912186

Williams, Sarah, "Ghost Cities of China—Civic Data Design Lab," n.d.
https://civicdatadesignlab.mit.edu/Ghost-Cities-of-China

Financial Post, "Does Everyone Have the Chinese 'Ghost Towns' Story All Wrong?," March 13, 2013
https://financialpost.com/business-insider/china-ghost-towns.

Powell, Bill, "Ghost City," April 5, 2010
http://www.time.com/time/magazine/article/0%2C9171%2C1975336%2C00.html

Sheehan, Matt, "Signs of Life In China's Gleaming 'Ghost City' Of Ordos." HuffPost, December 7, 2017
https://www.huffpost.com/entry/china-ordos-ghost-city-life_n_7204016

NOTES AND SOURCES

CRYPTOCURRENCY—THE REEXAMINATION OF MONEY

Nakamoto, Satoshi, "Bitcoin: A Peer-to-Peer Electronic Cash System"
https://bitcoin.org/bitcoin.pdf

Zeder, Raphael, "The Four Different Types of Money." Quickonomics, January 14, 2023
https://quickonomics.com/different-types-of-money/

Greenberg, Andy, "Crypto Currency" Forbes, April 20, 2011
https://www.forbes.com/forbes/2011/0509/technology-psilocybin-bitcoins-gavin-andresen-crypto-currency.html?sh=1230c148353e

MercoPress, "Bitcoin Legal Tender in El Salvador, First Country Ever," June 10, 2021
https://en.mercopress.com/2021/06/10/bitcoin-legal-tender-in-el-salvador-first-country-ever

Khalili, Joel, "The Fallout of the FTX Collapse," WIRED, November 11, 2022
https://www.wired.com/story/the-fallout-of-the-ftx-collapse/

Davies, Pascale, "Bitcoin Mining Is Actually Worse for the Environment since China Banned It, a New Study Says," Euronews, February 26, 2022
https://www.euronews.com/next/2022/02/26/bitcoin-mining-was-actually-worse-for-the-environment-since-china-banned-it-a-new-study-sa

Hutcheon, Stephen, "The Rise and Rise of Dogecoin, the Internet's Hottest Cryptocurrency," The Sydney Morning Herald, January 24, 2014
https://www.smh.com.au/technology/the-rise-and-rise-of-dogecoin-the-internets-hottest-cryptocurrency-20140124-31d24.html

Ball, James, "Silk Road: The Online Drug Marketplace That Officials Seem Powerless to Stop," The Guardian, December 1, 2017
https://www.theguardian.com/world/2013/mar/22/silk-road-online-drug-marketplace

Mullin, Joe, "Ulbricht Guilty in Silk Road Online Drug-Trafficking Trial," Ars Technica, February 5, 2015
https://arstechnica.com/tech-policy/2015/02/ulbricht-guilty-in-silk-road-online-drug-trafficking-trial/

Gamboa, Glenn, and Daniel, Will, "'Literally, There's No Record-Keeping Whatsoever': FTX's New CEO Is Flabbergasted, and D.C. Is Laughing at SBF Using QuickBooks," Fortune, December 14, 2022
https://fortune.com/2022/12/13/ftx-ceo-john-ray-testifies-congress-no-record-keeping-quickbooks-bankman-fried/

Vanguard X, "Learn the Story of the Programmer Who Lost Millions in Bitcoin," June 1, 2023
https://vanguard-x.com/blockchain/lost-millions-in-bitcoin/

Shumba, Camomile, "Michael Saylor's MicroStrategy Buys Another 7,002 Bitcoins for $414 Million—Keeping Its Promise to Keep Adding the Crypto," Markets Insider, November 29, 2021
https://markets.businessinsider.com/news/currencies/michael-saylor-microstrategy-crypto-bitcoin-buy-holdings-sell-stock-2021-11

NOTES AND SOURCES

THE BERNIE MADOFF PONZI SCHEME

U.S. Securities and Exchange Commission Office of Investigations, "Investigation of Failure of the SEC to Uncover Bernard Madoff's Ponzi Scheme—Public Version -." www.sec.gov, August 9, 2009
https://www.sec.gov/news/studies/2009/oig-509.pdf.

"Biggest Fraud in History $50 Billion Madoff Ponzi Scheme: The Market Oracle"
http://www.marketoracle.co.uk/Article7769.html.

Markopolos, Harry, No One Would Listen, John Wiley & Sons, 2011, ISBN 978-0-470-55373-2

Dow Jones Reuters Business Interactive LLC, "Don't Ask, Don't Tell: Bernie Madoff Is so Secretive, He Even Asks His Investors to Keep Mum," U.S. Securities and Exchange Commission, December 11, 2003
https://www.sec.gov/news/studies/2009/oig-509/exhibit-0156.pdf.

Henriques, Diana B, The Wizard of Lies. Macmillan, 2011, ISBN 9781250116581

WHOLE LOTTO DELUSION

"A Brief History of Lotteries Around the World" The Lottery House, July 21, 2023
https://thelotteryhouse.com/blog/a-brief-history-of-lotteries-around-the-world

"The Development of Ancient China and Lotteries," The Lottery House, August 17, 2023
https://thelotteryhouse.com/blog/the-development-of-ancient-china-and-lotteries

"EuroMillions Lottery Scams—How to Detect a EuroMillions Scam,"
https://www.euro-millions.com/scams

"Fake Prize, Sweepstakes, and Lottery Scams," February 9, 2022
https://consumer.ftc.gov/articles/fake-prize-sweepstakes-lottery-scams

"Lottery Mathematics," August 21, 2023
https://en.wikipedia.org/wiki/Lottery_mathematics

"How Hard Is It to Win the Lottery? Odds to Keep in Mind as Powerball and Mega Millions Jackpots Soar, AP News," July 19, 2023
https://apnews.com/article/powerball-mega-millions-winning-odds-numbers-a3e5a8e8e7ed15d7500c1d6acdab6785

MULTI-LEVEL MARKETING

Federal Trade Commission, "Multi-Level Marketing Businesses and Pyramid Schemes," November 25, 2022
https://consumer.ftc.gov/articles/multi-level-marketing-businesses-pyramid-schemes

Karp, Gregory, "The Fine Line between Legitimate Businesses and Pyramid Schemes," Chicago Tribune, February 10, 2013
https://www.chicagotribune.com/business/ct-xpm-2013-02-10-ct-biz-0210-herbalife-20130210-story.html

NOTES AND SOURCES

Federal Trade Commission, "FTC Sends Warning Letters to Multi-Level Marketers Regarding Health and Earnings Claims They or Their Participants Are Making Related to Coronavirus," April 11, 2023
https://www.ftc.gov/news-events/news/press-releases/2020/04/ftc-sends-warning-letters-multi-level-marketers-regarding-health-earnings-claims-they-or-their

Federal Trade Commission, "FTC Again Warns Multi-Level Marketers about Unproven Health and Earnings Claims," June 13, 2022
https://www.ftc.gov/business-guidance/blog/2020/06/ftc-again-warns-multi-level-marketers-about-unproven-health-and-earnings-claims

"Income Disclosure"
https://www.amway.com/en_US/income-disclosure

Taylor, Ph.D., MBA, Jon M., "The Case (For And) Against Multi-Level Marketing," centerforinquiry.org
https://centerforinquiry.org/wp-content/uploads/sites/33/quackwatch/taylor.pdf

Robert Todd Carroll, "Multi-Level Marketing (a.k.a. Network Marketing & Referral Marketing), The Skeptic's Dictionary—Skepdic.Com, https://www.skepdic.com/mlm.html

Brittney Laryea, "Survey: Vast Majority of Multilevel Marketing Participants Earn Less Than 70 Cents an Hour," September 17, 2018, https://www.magnifymoney.com/news/mlm-participants-survey/

TIMESHARES

Molina, Linda, "Timeshare Salespeople—Who Is In On The SCAM?" September 14, 2023
https://www.timesharescam.com/blog/77-timeshare-salespeople/

Federal Trade Commission, "Timeshares, Vacation Clubs, and Related Scams," July 27, 2021
https://consumer.ftc.gov/articles/timeshares-vacation-clubs-related-scams

Max, Sarah, "The Timeshare Trap," March 21, 2002
http://money.cnn.com/2002/03/21/pf/yourhome/q_timeshare/

Elliott, Christopher, "Trapped in a Timeshare? Here's How to Escape," USA TODAY, December 26, 2018
https://www.usatoday.com/story/travel/advice/2018/12/26/timeshare-troubles-extricate-unwanted-unit/2375107002/

"U.S. Timeshare Industry: By the Numbers," ARDA,
https://www.arda.org/news-communications/timeshare-industry-basics/us-timeshare-industry-numbers

"Timeshare Industry Statistics And Trends in 2023," September 5, 2023
https://blog.gitnux.com/timeshare-industry-statistics/

Federal Trade Commission, "FTC and Wisconsin Aim to Show Deceptive Timeshare Exit Claims the Exit," November 23, 2022
https://www.ftc.gov/business-guidance/blog/2022/11/ftc-wisconsin-aim-show-deceptive-timeshare-exit-claims-exit

Federal Trade Commission, "Be on the Lookout for Timeshare Resale Phonies," May 17, 2022
https://consumer.ftc.gov/consumer-alerts/2014/05/be-lookout-timeshare-resale-phonies

"Learn Why Timeshare Math Doesn't Add Up," https://blog.canceltimeshare.io/timeshare-math.

NOTES AND SOURCES

VOLUME 3—YOUR SOUL (CULTURAL, RELIGIOUS, AND POLITICAL DELUSIONS)

CONSPIRACY THEORIES—AN OVERVIEW

Neuroscience News, "The Psychology of Conspiracy Theorists: More Than Just Paranoia," June 26, 2023
https://neurosciencenews.com/psychology-conspiracy-theories-23531/

Jolley, Daniel, and Karen M. Douglas, "The Effects of Anti-Vaccine Conspiracy Theories on Vaccination Intentions," Public Library of Science, February 20, 2014
https://doi.org/10.1371/journal.pone.0089177

Cassam, Quassim, "The Intellectual Character of Conspiracy Theorists," Aeon, August 3, 2020
https://aeon.co/essays/the-intellectual-character-of-conspiracy-theorists

Shermer, Michael, "The Conspiracy Theory Detector." Scientific American. Nature Portfolio, December 1, 2010
https://doi.org/10.1038/scientificamerican1210-102

McKeown, Trevor W., "A Bavarian Illuminati Primer," Grand Lodge of British Columbia and Yukon
https://freemasonry.bcy.ca/texts/illuminati.html

THE MOTHER OF ALL CONSPIRACY THEORIES

National Archives, "Warren Commission Report"
https://www.archives.gov/research/jfk/warren-commission-report

National Archives. "House Select Committee on Assassinations"
https://www.archives.gov/research/jfk/select-committee-report

SOVEREIGN CITIZEN MOVEMENT

"Sovereign Citizens Movement," Southern Poverty Law Center
https://www.splcenter.org/fighting-hate/extremist-files/ideology/sovereign-citizens-movement

"The Sovereign Citizen Movement," Federal Bureau of Investigation
https://archives.fbi.gov/archives/news/stories/2010/april/sovereigncitizens_041310/domestic-terrorism-the-sovereign-citizen-movement

"A Quick Guide To Sovereign Citizens," soc.unc.edu (UNC School of Government), November 2013
https://www.sog.unc.edu/sites/www.sog.unc.edu/files/Sov%20citizens%20quick%20guide%20Nov%202013.pdf

Rooke, Associate Chief Justice J.D., "Meads v. Meads, 2012 ABQB 571 (CanLII)," canlii.org, September 18, 2012
https://www.canlii.org/en/ab/abqb/doc/2012/2012abqb571/2012abqb571.html.

HOLOCAUST DENIAL

United States Holocaust National Museum, "Holocaust Deniers and Public Misinformation"
https://encyclopedia.ushmm.org/content/en/article/holocaust-deniers-and-public-misinformation

Novella, Steven, "Holocaust Denial," The NESS—New England Skeptical Society, July 19, 2009
https://theness.com/index.php/holocaust-denial/

Pengelly, Martin, and Joan E Greve, "Fury as Marjorie Taylor Greene Likens Covid Rules to Nazi Treatment of Jews," The Guardian, May 25, 2021
https://www.theguardian.com/us-news/2021/may/25/marjorie-taylor-greene-nazi-jews-condemnation

Steele, Alistair, "Disgust Growing over Vaccine Protesters' Holocaust Comparisons," CBC, September 15, 2021
https://www.cbc.ca/news/canada/ottawa/vaccine-protesters-holocaust-comparisons-1.6175321

UN News "UN General Assembly Approves Resolution Condemning Holocaust Denial," January 25, 2022
https://news.un.org/en/story/2022/01/1110202

UNIDENTIFIED FLYING OBJECTS

Wright, Phil, "Flying Saucers Still Evasive 70 Years after Pilot's Report." Spokesman.com, June 25, 2017
https://www.spokesman.com/stories/2017/jun/25/flying-saucers-still-evasive-70-years-after-pilots/

Office of the Director of National Intelligence, "Preliminary Assessment: Unidentified Aerial Phenomena," June 25, 2021
https://www.dni.gov/files/ODNI/documents/assessments/Prelimary-Assessment-UAP-20210625.pdf

Richwine, Lisa, "Area 51 Raid Lures Festive UFO Hunters to Nevada Desert; Five Arrested," Reuters, U.S., September 21, 2019
https://www.reuters.com/article/us-usa-area51/in-nevada-desert-area-51-raid-lures-festive-ufo-hunters-three-arrested-idUSKBN1W51H6

Griffin, Andrew, "Storm Area 51: Are Alien-Hunters Really Planning to 'Raid' the Secret US Military Base?," The Independent, July 16, 2019
https://www.independent.co.uk/tech/storm-area-51-when-where-aliens-military-base-raid-facebook-event-a9005546.html

Arronz, Adolfo, and Pablo Robles, "UFO Sightings." South China Morning Post, Accessed October 2, 2023
https://multimedia.scmp.com/culture/article/ufo/index.html

"National UFO Reporting Center"
https://nuforc.org/

Dunning, Brian, "Aliens in Roswell," Skeptoid, December 18, 2007
https://skeptoid.com/episodes/4079

Mitsanas, Michael, "Here Are the 5 Most Memorable Moments from Congress' UFO Hearing," NBC News, July 26, 2023
https://www.nbcnews.com/politics/congress/are-5-memorable-moments-congress-ufo-hearing-rcna96476

NOTES AND SOURCES

FLAT EARTH

Brazil, Rachel, "Fighting Flat-Earth Theory," Physics World, July, 2020
https://physicsworld.com/a/fighting-flat-earth-theory/

McIntyre, Lee, "Flat Earthers, and the Rise of Science Denial in America, Newsweek, May 14, 2019
https://www.newsweek.com/flat-earth-science-denial-america-1421936

"Infamous Daredevil Mad Mike Hughes Has Died in Homemade Rocket Crash in California," ScienceAlert, February 23, 2020
https://www.sciencealert.com/infamous-flat-earth-daredevil-dies-in-crash-in-california

Dubay, Eric, "200 Proofs Earth Is Not a Spinning Ball!.Pdf"
https://docs.google.com/file/d/0B5Dy_Ci78cCvazRqdFZoTUVyN2M/preview?resourcekey=0-TbM3b4oNfPeUXewomBG2mQ

HIV/AIDS DENIALISM

"HIV/AIDs Timeline," Centers for Disease Control, National Prevention Information Network
https://npin.cdc.gov/pages/hiv-and-aids-timeline#1980

Boseley, Sarah, "Discredited Doctor's 'cure' for Aids Ignites Life-and-Death Struggle in South Africa." the Guardian, May 14, 2005
https://www.theguardian.com/world/2005/may/14/southafrica.internationalaidanddevelopment

"AIDS Denialists Who Have Died'" AIDSTruth.Org,
https://web.archive.org/web/20100126000256/http://www.aidstruth.org/denialism/dead_denialists

Nattrass, Nicoli, "AIDS Denialism vs. Science," Skeptical Inquirer, September / October, 2007
https://skepticalinquirer.org/2007/09/aids-denialism-vs-science/

"The Politics of HIV/AIDS in South Africa," Journ-AIDS
https://web.archive.org/web/20070702083352/http://journaids.org/politicsofhiv.php

"Who Was Ryan White?," Ryan White HIV/AIDS Program," February, 2022
https://ryanwhite.hrsa.gov/about/ryan-white

"Pope Rejects Condoms for Africa," BBC NEWS Europe, June 10, 2005
http://news.bbc.co.uk/2/hi/europe/4081276.stm

Wu, Zunyou, Chen Jun-Fang, Sarah Robbins Scott, and Jennifer M. McGoogan, "History of the HIV Epidemic in China," Current Hiv/aids Reports, Springer Science+Business Media, November 26, 2019
https://doi.org/10.1007/s11904-019-00471-4

CLIMATE CHANGE DENIALISM

Lynas, Mark, Benjamin Z Houlton and Simon Perry, Environmental Research Letters, Volume 16, Number 11. "Greater than 99% Consensus on Human Caused Climate Change in the Peer-Reviewed Scientific Literature"
https://iopscience.iop.org/article/10.1088/1748-9326/ac2966

NOTES AND SOURCES

Young, Élan, "Coal Knew, Too," HuffPost, December 16, 2019
https://www.huffpost.com/entry/coal-industry-climate-change_n_5dd6bbebe4b0e29d7280984f

The Climate Reality Project, "The Climate Denial Machine: How the Fossil Fuel Industry Blocks Climate Action," September 5, 2019
https://www.climaterealityproject.org/blog/climate-denial-machine-how-fossil-fuel-industry-blocks-climate-action

CORONAVIRUS CONSPIRACY THEORIES

Imhoff, Roland, and Pia Lamberty, "A Bioweapon or a Hoax? The Link Between Distinct Conspiracy Beliefs About the Coronavirus Disease (COVID-19) Outbreak and Pandemic Behavior," Social Psychological and Personality Science, SAGE Publishing, July 6, 2020
https://doi.org/10.1177/1948550620934692

    Pradelle, Alexiane, Sabine Mainbourg, Steeve Provencher, E. Massy, Guillaume Grenet, and J.C. Lega, "Deaths Induced by Compassionate Use of Hydroxychloroquine during the First COVID-19 Wave: An Estimate." Biomedicine & Pharmacotherapy, February 1, 2024, https://doi.org/10.1016/j.biopha.2023.116055

Lynas, Mark, "COVID: Top 10 Current Conspiracy Theories" Alliance for Science," April 20, 2020
https://allianceforscience.org/blog/2020/04/covid-top-10-current-conspiracy-theories/

Polidoro, Massimo, "Stop the Epidemic of Lies! Thinking about COVID-19 Misinformation." Skeptical Inquirer, July / August, 2020
https://skepticalinquirer.org/2020/06/stop-the-epidemic-of-lies-thinking-about-covid-19-misinformation/

Broderick, Ryan, "QAnon Supporters, Anti-Vaxxers Spread A Hoax Bill Gates Created Coronavirus." BuzzFeed News, January 23, 2020
https://www.buzzfeednews.com/article/ryanhatesthis/QAnon-supporters-and-anti-vaxxers-are-spreading-a-hoax-that

Wynne, Kelly, "YouTube Video Suggests 5G Internet Causes Coronavirus and People Are Falling for It," Newsweek, March 19, 2020
https://www.newsweek.com/youtube-video-suggests-5g-internet-causes-coronavirus-people-are-falling-it-1493321

"Mast Fires Surge in the UK over Easter Weekend amid 5G-Coronavirus Conspiracy Theories," Irish Examiner, April 14, 2020
https://www.irishexaminer.com/world/arid-30994035.html

Jingnan, Huo, "Why There Are So Many Different Guidelines For Face Masks For The Public." NPR, April 10, 2020
https://www.npr.org/sections/goatsandsoda/2020/04/10/829890635/why-there-so-many-different-guidelines-for-face-masks-for-the-public

Hannon, Elliot, "Hundreds Die in Iran From Bootleg Alcohol Being Peddled Online as Fake Coronavirus Remedy" Slate Magazine, March 27, 2020
https://slate.com/news-and-politics/2020/03/hundreds-die-iran-drinking-bootleg-alcohol-methanol-coronavirus-cure-social-media.html

Robins-Early, Nick, "The Strange Origins Of Trump's Hydroxychloroquine Obsession," HuffPost, May 13, 2020
https://www.huffpost.com/entry/trump-hydroxychloroquine-coronavirus-fox-news_n_5ebaffdbc5b65b5fd63dac80

"WHO Advises That Ivermectin Only Be Used to Treat COVID-19 within Clinical Trials," World Health Organization, March 31, 2021,
https://www.who.int/news-room/feature-stories/detail/who-advises-that-ivermectin-only-be-used-to-treat-covid-19-within-clinical-trials

Schraer, Rachel and Jack Goodman, "Ivermectin: How False Science Created a Covid 'miracle' Drug," BBC News, October 6, 2021
https://www.bbc.com/news/health-58170809

Gallagher, Ryan, "5G Coronavirus Conspiracy Theory Driven by Coordinated Effort," Al Jazeera, April 10, 2020
https://www.aljazeera.com/economy/2020/4/10/5g-coronavirus-conspiracy-theory-driven-by-coordinated-effort

Sommerlad, Joe, "What Is Reddit's Herman Cain Award?" The Independent, September 29, 2021
https://www.independent.co.uk/tech/herman-cain-award-reddit-anti-vax-b1929219.html

YEAR 2000 BUG (Y2K)

BBC NEWS, Asia-Pacific, Japan Nuclear Plants Malfunction
http://news.bbc.co.uk/2/hi/asia-pacific/585950.stm

BBC News, TALKING POINT, Y2K: Overhyped and Oversold?"
http://news.bbc.co.uk/2/hi/talking_point/586938.stm

The Globe and Mail, "Opinion: The Y2K Bug Turned out to Be a Non-Event, Eric Reguly Says—The Globe and Mail"
https://www.theglobeandmail.com/amp/report-on-business/rob-commentary/the-y2k-bug-turned-out-to-be-a-non-event-eric-reguly-says/article765124/

Gibbs, Thom, "The Millennium Bug Myth, 20 Years on: Why You're Probably Wrong about Y2K." The Telegraph, December 19, 2019
https://www.telegraph.co.uk/technology/2019/12/19/millennium-bug-myth-20-years-probably-wrong-y2k/

Washingtonpost.com, Business Y2K Computer Bug
https://www.washingtonpost.com/wp-srv/business/longterm/y2k/stories/consumer_faith.htm

Rifkin, Ira, "NEWS FEATURE: Christian Conservatives Reconsider Y2K Disaster Warnings." Religion News Service, October 27, 1999
https://religionnews.com/1999/10/27/news-feature-christian-conservatives-reconsider-y2k-disaster-warnings/

Time.com, "TIME Magazine Cover: End of the World—Jan. 18, 1999," January 18, 1999
https://content.time.com/time/covers/0,16641,19990118,00.html

NOTES AND SOURCES

NESSIE, BIGFOOT, AND THE ABOMINABLE SNOWMAN

"The Loch Ness Centre"
   https://lochness.com/

"The Surgeon's Hoax—Loch Ness Hoax Photo," The Museum of Unnatural History
   http://www.unmuseum.org/nesshoax.htm

Edwards, Phil, "How Scientists Debunked the Loch Ness Monster." Vox, October 20, 2015
   https://www.vox.com/2015/4/21/8459353/loch-ness-monster

Stagnaro, Angelo, "St. Columba and the Loch Ness Monster," National Catholic Register, November 25, 2018
   https://www.ncregister.com/blog/st-columba-and-the-loch-ness-monster

"Legend of Nessie—Ultimate and Official Loch Ness Monster Site—Searching for Nessie," Legend of Nessie,
   http://www.nessie.co.uk/htm/the_nessie_hunters/hunter1.html

"Hundreds Join Largest Loch Ness Monster Hunt in 50 Years in Scotland," Reuters, August 27, 2023
   https://www.reuters.com/world/uk/hundreds-join-largest-loch-ness-monster-hunt-50-years-scotland-2023-08-27/

Crair, Ben, "Why Do So Many People Still Want to Believe in Bigfoot?" Smithsonian Magazine, August 21, 2018
   https://www.smithsonianmag.com/history/why-so-many-people-still-believe-in-bigfoot-180970045/

Flight, Tim, "The Hairy History of Bigfoot in 20 Intriguing Events," History Collection, November 9, 2018
   https://historycollection.com/the-hairy-history-of-bigfoot-in-20-intriguing-events/10/

McKenzie, Steven, "Scientists Challenge 'Abominable Snowman DNA' Results," BBC News, December 17, 2014
   https://www.bbc.com/news/uk-scotland-highlands-islands-30479718

Ghosh, Devarsi, "A Short History of Yeti Mania, from the Ramsay Brothers to Tintin," Scroll.in, May 1, 2019
   https://scroll.in/article/921783/a-short-history-of-yeti-mania-from-the-ramsay-brothers-to-tintin

CHEMTRAILS

Suzuki, David, "Conspiracy Theories Fuel Chemtrail Beliefs and Climate Change Denial," Earth Island Journal, September 5, 2013
   https://www.earthisland.org/journal/index.php/articles/entry/conspiracy_theories_fuel_climate_change_denial_and_chemtrail_beliefs/

Keith, David, "Chemtrails Conspiracy Theory," The Keith Group,
   https://keith.seas.harvard.edu/chemtrails-conspiracy-theory

Dunne, Carey, "My Month with Chemtrails Conspiracy Theorists," the Guardian, December 19, 2017
   https://www.theguardian.com/environment/2017/may/22/california-conspiracy-theorist-farmers-chemtrails

Newitz, Annalee and Adam Steiner, "Here's Where the Chemtrail Conspiracy Theory Actually Came From," Gizmodo, December 16, 2015
https://gizmodo.com/is-that-reflective-cloud-about-to-poison-you-and-change-1638680856

Coleman, Alistair, "Chemtrails: What's the Truth behind the Conspiracy Theory?" BBC News, July 22, 2022
https://www.bbc.com/news/blogs-trending-62240071

THE BIRTHER MOVEMENT

"Certificate of Live Birth," National Archives, August 4, 1961
https://obamawhitehouse.archives.gov/sites/default/files/rss_viewer/birth-certificate.pdf

Jardina, Ashley, and Michael W. Traugott, "The Genesis of the Birther Rumor: Partisanship, Racial Attitudes, and Political Knowledge," Journal of Race, Ethnicity, and Politics, Cambridge University Press, November 20, 2018
https://www.cambridge.org/core/journals/journal-of-race-ethnicity-and-politics/article/genesis-of-the-birther-rumor-partisanship-racial-attitudes-and-political-knowledge/8C13EDF7D45A475E97B5D2B35BC8979E#

Stableford, Dylan, "'Born in Kenya': Obama's Literary Agent Misidentified His Birthplace in 1991," ABC News, May 18, 2012
https://web.archive.org/web/20120706093227/https://abcnews.go.com/Politics/OTUS/born-kenya-obamas-literary-agent-misidentified-birthplace-1991/story?id=16372566#.Vj6h4r9mqjc

Koppelman, Alex, "No, Obama's Grandmother Didn't Say He Was Born in Kenya," Salon, July 24, 2009
https://www.salon.com/2009/07/24/liddy/

Koppelman, Alex, "Source for Forged Kenyan Birth Certificate Found?" Salon, August 4, 2009
https://www.salon.com/2009/08/04/australia_certificate/

PSYCHIC SERVICES

Shermer, Michael, "Science Friction: Excerpt; Chapter One—Psychic for a Day (or How I Learned Tarot Cards, Palm Reading, Astrology, and Mediumship in 24 Hours)," June 22, 2020
https://michaelshermer.com/science-friction/excerpt/

Jarry, Jonathan, McGill Office for Science and Society, "How Astrology Escaped the Pull of Science," October 9, 2020
https://www.mcgill.ca/oss/article/pseudoscience/how-astrology-escaped-pull-science

Gecewicz, Claire, Pew Research Center, "'New Age' Beliefs Common among Religious, Nonreligious Americans," October 1, 2018
https://www.pewresearch.org/short-reads/2018/10/01/new-age-beliefs-common-among-both-religious-and-nonreligious-americans/

Carroll, Robert Todd, "Divination—Fortune Telling—The Skeptic's Dictionary—Skepdic.Com"
https://skepdic.com/divinati.html

Carroll, Robert Todd, "Palmistry—The Skeptic's Dictionary—Skepdic.Com"
https://skepdic.com/palmist.html

NOTES AND SOURCES

Rickert, Philip V., "Confessions of a Palmist | News | The Harvard Crimson," January 10, 1963
https://www.thecrimson.com/article/1968/1/10/confessions-of-a-palmist-pithe-name/

AARP, "What You Need to Know About Psychic Scams," August 31, 2023
https://www.aarp.org/money/scams-fraud/info-2022/psychic.html

Carroll, Robert Todd, "Parapsychology—The Skeptic's Dictionary—Skepdic.Com"
https://www.skepdic.com/parapsy.html.

Reich, Aaron, "Uri Geller Museum: Mind, Life, Legacy of Israel's Most Interesting Man," The Jerusalem Post, October 23, 2021
https://www.jpost.com/israel-news/culture/uri-geller-museum-a-look-into-the-mind-life-legacy-of-israels-most-interesting-personality-682600

"Did Psychic Jeane Dixon Predict JFK's Assassination?," The Straight Dope, February 2, 2000
https://www.straightdope.com/21342875/did-psychic-jeane-dixon-predict-jfk-s-assassination

Kohn, David, "Shirley MacLaine's Recent Lives," CBS News, September 18, 2002
https://www.cbsnews.com/news/shirley-maclaines-recent-lives/

ALEX JONES AND THE SANDY HOOK SCHOOL SHOOTING

McGovern, Tim, "Far-Right Personality Alex Jones Banned from Facebook & Instagram for Being a 'Dangerous Individual,'" Peoplemag, May 3, 2019
https://people.com/human-interest/alex-jones-banned-facebook-instagram/

Squire, Megan and Michael Edison Hayden, "'Absolutely Bonkers': Inside Infowars' Money Machine," Southern Poverty Law Center, March 8, 2023
https://www.splcenter.org/hatewatch/2023/03/08/absolutely-bonkers-inside-infowars-money-machine

Johnson, Timothy, "Trump Ally Alex Jones Doubles Down On Sandy Hook Conspiracy Theories," Media Matters for America, November 17, 2016
https://www.mediamatters.org/donald-trump/trump-ally-alex-jones-doubles-down-sandy-hook-conspiracy-theories

Collins, Dave, "Alex Jones Ordered to Pay $965 Million for Sandy Hook Lies" AP News, October 14, 2022
https://apnews.com/article/shootings-school-connecticut-conspiracy-alex-jones-3f579380515fdd6eb59f5bf0e3e1c08f

Warzel, Charlie, "Alex Jones Just Can't Help Himself," BuzzFeed News, May 4, 2017
https://www.buzzfeednews.com/article/charliewarzel/alex-jones-will-never-stop-being-alex-jones

Murdock, Sebastian, "Alex Jones' Infowars Store Made $165 Million Over 3 Years, Records Show," HuffPost, January 7, 2022
https://www.huffpost.com/entry/infowars-store-alex-jones_n_61d71d8fe4b0bcd2195c6562

Murdock, Sebastian, "FDA To Alex Jones: Stop Selling Fake Coronavirus Cures," HuffPost, April 11, 2020
https://www.huffpost.com/entry/fda-to-alex-jones-stop-selling-fake-coronavirus-cures_n_5e91e4a8c5b69d65062a09da

NOTES AND SOURCES

Higgins, Tucker, "Alex Jones' 5 Most Disturbing and Ridiculous Conspiracy Theories," CNBC, September 15, 2018
https://www.cnbc.com/2018/09/14/alex-jones-5-most-disturbing-ridiculous-conspiracy-theories.html

Sommer, Will, "InfoWars' Alex Jones Accused of 'Jaw-Dropping' Scheme to Hide Money From Sandy Hook Families in Texas Lawsuit," The Daily Beast, April 8, 2022
https://www.thedailybeast.com/infowars-alex-jones-accused-of-jaw-dropping-scheme-to-hide-money-from-sandy-hook-families-in-texas-lawsuit?ref=author

Nguyen, Alex, "Alex Jones Files for Bankruptcy after Sandy Hook Judgments" The Texas Tribune, December 2, 2022
https://www.texastribune.org/2022/12/02/alex-jones-bankruptcy/

CATFISHING

"What Is Catfishing Online: Signs & How to Tell," Fortinet,
https://www.fortinet.com/resources/cyberglossary/catfishing#:~:text=Catfishing%20refers%20to%20when%20a,that%20it%20is%20their%20own

Zimmer, Ben, "Catfish: How Manti Te'o's Imaginary Romance Got Its Name," BostonGlobe.com, January 27, 2013
https://www.bostonglobe.com/ideas/2013/01/27/catfish-how-manti-imaginary-romance-got-its-name/inqu9zV8RQ7j19BRGQkH7H/story.html

"Catfishing"
https://www.cybersmile.org/what-we-do/advice-help/catfishing

D'Costa, Krystal, "Catfishing: The Truth About Deception Online." Scientific American Blog Network, April 25, 2014
https://blogs.scientificamerican.com/anthropology-in-practice/catfishing-the-truth-about-deception-online/

CONSPIRACY THEORY SATIRE—BIRDS AREN'T REAL

Alfonsi, Sharyn, "Parodying Conspiracy Theories with the Birds Aren't Real Movement," 60 Minutes, CBS News, May 2, 2022
https://www.cbsnews.com/news/birds-arent-real-peter-mcindoe-60-minutes-2022-05-01/

Koch, Mitchell , "Every Tweet Is a Lie: Birds Aren't Real Campaign Spreads Message with New Memphis Billboard," WREG Memphis News Chanel 3, July 18, 2019
https://wreg.com/news/every-tweet-is-a-lie-birds-arent-real-campaign-spreads-message-with-new-memphis-billboard/

Palma, Bethania, "What Is the 'Birds Aren't Real' Movement?" Snopes, November 22, 2021
https://www.snopes.com/articles/381287/birds-arent-real-movement/

Alfonso, Fernando, "Are Birds Actually Government-Issued Drones? So Says a New Conspiracy Theory Making Waves (and Money)," Audubon, November 16, 2018
https://www.audubon.org/news/are-birds-actually-government-issued-drones-so-says-new-conspiracy-theory-making

Lorenz, Taylor "Birds Aren't Real, or Are They? Inside a Gen Z Conspiracy Theory," The New York Times, December 9, 2021,
https://www.nytimes.com/2021/12/09/technology/birds-arent-real-gen-z-misinformation.html

RELIGIOUS DELUSIONS

MAINSTREAM RELIGION

"The Global Religious Landscape," Pew Research Center's Religion & Public Life Project, December 18, 2012
https://www.pewresearch.org/religion/2012/12/18/global-religious-landscape-exec/

TELEVANGELISM

Falsani, Cathleen, "The Worst Ideas of the Decade," Washington Post
https://www.washingtonpost.com/wp-srv/special/opinions/outlook/worst-ideas/prosperity-gospel.html

"The Bankruptcy of the Prosperity Gospel: An Exercise in Biblical and Theological Ethics," Bible.Org,
https://bible.org/article/bankruptcy-prosperity-gospel-exercise-biblical-and-theological-ethics

Sherman, Bill, "Special Section—1918—2009 Oral Roberts Legacy," Tulsa World, December 20, 2009
https://tulsaworld.com/app/oralroberts/pdf/specialsection.pdf

Harris, Art, "Robertson's Bakker Connection," Washington Post, February 6, 1988
https://www.washingtonpost.com/archive/lifestyle/1988/02/06/robertsons-bakker-connection/f558f67c-c4f5-489c-b733-e768d1daacdc/

Isikoff, Michael, "PTL FUND RAISING A TANGLED SAGA," Washington Post, May 23, 1987
https://www.washingtonpost.com/archive/politics/1987/05/23/ptl-fund-raising-a-tangled-saga/e659bb79-63e3-4a98-bfc6-c27dfec4fc07/?noredirect=on

Wigger, John, "Prosperity Gospel Apocalypse: Jim and Tammy Faye Bakker's PTL Empire" Anxious Bench, July 26, 2017
https://www.patheos.com/blogs/anxiousbench/2017/07/sex-money-fame-and-the-evangelical-empire-of-jim-and-tammy-faye-bakker/

McKinney, Kelsey, "The Second Coming Of Televangelist Jim Bakker," BuzzFeed News, May 19, 2017
https://www.buzzfeednews.com/article/kelseymckinney/second-coming-of-televangelist-jim-bakker

Kuruvilla, Carol, "Televangelist Kenneth Copeland Defends His Private Jets: 'I'm A Very Wealthy Man.'" HuffPost, June 6, 2019
https://www.huffpost.com/entry/kenneth-copeland-jet-inside-edition_n_5cf822fee4b0e63eda94de4f

Lemon, Jason, "Conservative Pastor Claims He 'Healed' Viewers of Coronavirus Through Their TV Screens," Newsweek, March 12, 2020
https://www.newsweek.com/conservative-pastor-claims-he-healed-viewers-coronavirus-through-their-tv-screens-1492044

"The Fall of Jimmy Swaggart," Peoplemag, March 7, 1988
https://people.com/archive/cover-story-the-fall-of-jimmy-swaggart-vol-29-no-9/

Harris, Art, "JIMMY SWAGGART AND THE SNARE OF SIN." Washington Post, February 25, 1988
https://www.washingtonpost.com/archive/lifestyle/1988/02/25/jimmy-swaggart-and-the-snare-of-sin/d07127d2-c412-4738-98d9-3b186d1b92f9/

FAITH HEALING

Stephen Barrett, M.D., Quackwatch," December 27, 2009, Some Thoughts about Faith Healing,
https://quackwatch.org/related/faith/

Joe Nickell, Skeptical Inquirer, May/June 2002, "Benny Hinn: Healer or Hypnotist?—CSI"
https://web.archive.org/web/20131030140655/

http://www.csicop.org/si/show/benny_hinn_healer_or_hypnotist/

Carroll, Robert Todd, The Skeptic's Dictionary, "Faith Healing"
https://www.skepdic.com/faithhealing.html

Hall, Harriet, Science-Based Medicine, "Faith Healing," February 19, 2018
https://sciencebasedmedicine.org/faith-healing/

Maag, Christopher, Business Insider, "Scam Everlasting: After 25 Years, Debunked Faith Healer Still Preaching Debt Relief Scam," September 23, 2011
https://www.businessinsider.com/scam-everlasting-after-25-years-debunked-faith-healer-still-preaching-debt-relief-scam-2011-9?IR=T

Rita Swan, "Faith-Based Medical Neglect: For Providers and Policymakers," Springer Nature Switzerland, July 19, 2020
https://rdcu.be/dDx7C

SERPENT HANDLING

Wolf, John, Church Education Resource Ministries, "Snake Handlers"
https://web.archive.org/web/20160722140423/http://www.cerm.info/bible_studies/Apologetics/snake_handlers.htm

Cevallos, Danny, "Snakes and Church vs. State," May 28, 2014
https://www.cnn.com/2014/02/26/opinion/cevallos-snake-handling-law/index.html

CATHOLIC PRIEST SEX ABUSE COVER-UP

Thurston, Herbert, "Celibacy of the Clergy," The Catholic Encyclopedia, Vol. 3. New York: Robert Appleton Company, 1908
http://www.newadvent.org/cathen/03481a.htm

"StopBaptistPredators.Org"
https://stopbaptistpredators.org/index.htm

Corderus, Mike, News & World Report, "Dalai Lama Meets Alleged Victims of Abuse by Buddhist Gurus," September 14, 2018
https://www.usnews.com/news/world/articles/2018-09-14/dalai-lama-meets-alleged-victims-of-abuse-by-buddhist-gurus

Lewis, Aidan, BBC News, "Looking behind the Catholic Sex Abuse Scandal," May 4, 2010
http://news.bbc.co.uk/2/hi/8654789.stm

"Pope Ends 'secrecy' Rule on Child Sexual Abuse in Catholic Church," The Guardian, December 17, 2019
https://www.theguardian.com/world/2019/dec/17/pope-francis-ends-pontifical-secrecy-rule-child-sexual-abuse-catholic-church

John Jay College Of Criminal Justice, and Catholic Church. United States Conference Of Catholic Bishops, The nature and scope of sexual abuse of minors by Catholic priests and deacons in the United States, 2002: a research study conducted by the John Jay College of Criminal Justice, the City University of New York: for the United States Conference of Catholic Bishops. [Washington, D.C.: United States Conference of Catholic Bishops, ©, 2004] Pdf. Retrieved from the Library of Congress
www.loc.gov/item/2004303016/>

Blackwell, Eoin, "7 Percent Of All Catholic Priests Were Alleged Sex Abuse Perpetrators: Royal Commission." HuffPost, February 6, 2017
https://www.huffpost.com/archive/au/entry/catholic-church-under-royal-commission-spotlight_au_5cd39cb6e4b0ce845d82dcb4

John Jay College Of Criminal Justice, and Catholic Church. United States Conference Of Catholic Bishops, The Causes and Context of Sexual Abuse of Minors by Catholic Priests in the United States 1950-2010, May 2011
https://www.usccb.org/sites/default/files/issues-and-action/child-and-youth-protection/upload/The-Causes-and-Context-of-Sexual-Abuse-of-Minors-by-Catholic-Priests-in-the-United-States-1950-2010.pdf

Rezendes, Michael and Matt Carroll, Sacha Pfeiffer, and editor Walter V. Robinson, "Boston Globe / Spotlight / Abuse in the Catholic Church / The Geoghan Case," January 6, 2002
https://archive.boston.com/globe/spotlight/abuse/archive/stories/010602_geoghan.htm

"Collated USCCB Data On the Number of U.S. Priests Accused of Sexually Abusing Children and the Numbers of Persons Alleging Abuse 1950–2018"
https://www.bishop-accountability.org/AtAGlance/USCCB_Yearly_Data_on_Accused_Priests.htm

Thorp, Barbara, National Catholic Reporter, "20 Years after Boston Globe's 'Spotlight,' We Need a National Database of Accused Clergy," January 4, 2022
https://www.ncronline.org/news/accountability/20-years-after-boston-globes-spotlight-we-need-national-database-accused-clergy

CHURCH OF JESUS CHRIST OF LATTER-DAY SAINTS (MORMONISM)

Groberg, Lee, and Heidi Swinton and Gregory Peck, "American Prophet: The Story of Joseph Smith,"
https://www.pbs.org/americanprophet/joseph-smith.html

NOTES AND SOURCES

PBS, American Experience, "The Mormons," April 30, 2007
https://www.pbs.org/wgbh/americanexperience/films/mormons/

Embry, Jessie L., "Mormons," Encyclopedia.com, June 8, 2018
https://www.encyclopedia.com/philosophy-and-religion/christianity/protestant-denominations/mormons

Henderson, Peter, NBC News, "Mormon Church Earns $7 Billion a Year from Tithing, Analysis Indicates," August 13, 2012
https://www.nbcnews.com/news/investigations/mormon-church-earns-7-billion-year-tithing-analysis-indicates-flna939844

Curtis, Larry D., "LDS Church Releases Explanation of Its Use of Tithes, Donations after $100B Fund Revealed," KUTV, December 22, 2019
https://kutv.com/news/local/lds-church-releases-explanation-of-its-use-of-tithes-donations-after-100b-fund-revealed

Embry, Jessie L., "Mormons," Encyclopedia.com, June 8, 2018
https://www.encyclopedia.com/philosophy-and-religion/christianity/protestant-denominations/mormons

Noyce, David and Peggy Fletcher Stack, The Salt Lake Tribune, "LDS Church to Pay $5M for Hiding Stock Holdings, Needs to 'Rebuild Trust,'" February 23, 2023
https://www.sltrib.com/religion/2023/02/21/lds-church-investment-firm-agree/

Bloor, Steve, "The Futility and Brutality of Mormon Excommunication," Steve Bloor's Blog, November 11, 2015
https://stevebloor.wordpress.com/2015/11/11/the-futility-and-brutality-of-mormon-excommunication/

Harrison, Mette Ivie, "Do Mormons Shun?" HuffPost, November 6, 2017
https://www.huffpost.com/entry/do-mormons-shun_b_5a007e70e4b076eaaae27173

Henderson, Peter, NBC News, "Mormon Church Earns $7 Billion a Year from Tithing, Analysis Indicates," August 13, 2012
https://www.nbcnews.com/news/investigations/mormon-church-earns-7-billion-year-tithing-analysis-indicates-flna939844

JEHOVAH'S WITNESSES

"Jehovah's Witnesses—Official Website"
https://www.jw.org/en/

"BBC—Religions—Witnesses: Jehovah's Witnesses at a Glance," September 29, 2009
https://www.bbc.co.uk/religion/religions/witnesses/ataglance/glance.shtml

Woodward, Kenneth L., "Apocalypse Later." Newsweek, March 14, 2010
https://www.newsweek.com/apocalypse-later-180388

Dart, John, "Jehovah's Witnesses Abandon Key Tenet: Doctrine: Sect Has Quietly Retreated from Prediction That Those Alive in 1914 Would See End of World," Los Angeles Times, November 4, 1995
https://www.latimes.com/archives/la-xpm-1995-11-04-me-64883-story.html

NOTES AND SOURCES

Pew Research Center's Religion & Public Life Project, "Religion in America: U.S. Religious Data, Demographics and Statistics," June 13, 2022
https://www.pewresearch.org/religion/religious-landscape-study/religious-tradition/jehovahs-witness/

CHURCH OF ECKANKAR
"ECKANKAR, the Path of Spiritual Freedom," 2023
https://www.eckankar.org/

Tsakiris, Alex, "Dr. David Lane Not Sandbagged -- Patricia Churchland Part 2," Skeptiko—Science at the Tipping Point, May 14, 2018
https://skeptiko.com/240-david-lane-patricia-churchland-part-2/

Bellamy, Dodie, "Unauthorized Eckankar(Tm) Page," June 22, 1995
https://web.archive.org/web/20070703171210/http://www.geocities.com/eckcult/dodie.html

Lane, David Christopher (1994), The Making of a Spiritual Movement: The Untold Story of Paul Twitchell & Eckankar (revised ed.), Del Mar Press, ISBN 978-0-9611124-6-2
https://drive.google.com/file/d/1aSTd8LBJml86Z3yoz-ScyQ4UII6Er9Tq/view

DIANETICS AND THE CHURCH OF SCIENTOLOGY
Leiby, Richard, "SCIENTOLOGY FICTION," Washington Post, January 6, 2024. https://www.washingtonpost.com/archive/opinions/1994/12/25/scientology-fiction/809c906a-5145-4cce-a0fa-710d77adb5cd/

Corydon, Bent, and L. RON HUBBARD, JR., "L. RON HUBBARD Messiah or Madman?," Lyle Stuart Inc., 1987
https://web.archive.org/web/20130613033101/http://anonireland.com/content/wppdfcontent/books/messiahormadmen.pdf

Carroll, Robert Todd , "Dianetics—Scientology—The Skeptic's Dictionary" Skepdic.Com,
https://skepdic.com/dianetic.html

BEHARTIME, RICHARD, "Cover Story: The Thriving Cult of Greed and Power," Time.com, June 24, 2001
https://content.time.com/time/magazine/article/0,9171,156952,00.html

Owen, Chris, "Scientology Audited," November 3, 1997
https://www.cs.cmu.edu/~dst/Cowen/

Davies, Lizzy, "Church of Scientology Goes on Trial in France," the Guardian, May 25, 2009
https://www.theguardian.com/world/2009/may/25/scientology-france-fraud

Atack, Jon, "Scientology: Religion or Intelligence Agency?," October, 1995
https://home.snafu.de/tilman/j/berlin.html

"Court Order—FDA—Scientology Dianetics Hubbard E-Meter," Scribd.com March 1, 1973
https://www.scribd.com/document/5024758/Court-Order-FDA-Scientology-Dianetics-Hubbard-E-meter

NOTES AND SOURCES

Cavanaugh, Jeanne, "Scientology and the FDA: A Look Back, A Modern Analysis, And A New Approach," LEDA at Harvard Law School, April 7, 2004
https://dash.harvard.edu/bitstream/handle/1/8965552/Cavanaugh.html?sequence=2.

Mark, Michelle, "Lawsuits against the Church of Scientology Are Piling up, Alleging a Vast Network of Human Trafficking, Child Abuse, and Forced Labor," Insider, September 21, 2019
https://www.insider.com/scientology-lawsuits-allege-human-trafficking-forced-labor-child-abuse-2019-9

Kamer, Foster, "Scientology Leader David Miscavige: Still A Scary, Insane Psychopath." Gawker, August 2, 2009
https://www.gawker.com/5328446/scientology-leader-david-miscavige-still-a-scary-insane-psychopath?skyline=true&s=x

THE UNIFICATION CHURCH (MOONIES)

Davis, Freddy, "The Unification Church/Moonies," marketfaith.org, November 25, 2014
https://www.marketfaith.org/2014/11/the-unification-churchmoonies/

"Unification Church—Mass Moonie Marriage in the US," BBC News, November 29, 1997
http://news.bbc.co.uk/2/hi/special_report/1997/unification_church/34821.stm

McGill, Peter, "The Dark Shadow Cast by Moon Sun Myung's Unification Church and Abe Shinzo," The Asia-Pacific Journal: Japan Focus, October 15, 2022
https://apjjf.org/2022/17/McGill.html

END TIMES—THE MAYAN APOCALYPSE

Morrison, David, "Ask an Astrobiologist—Nibiru and Doomsday 2012: Questions and Answers," astrobiology2.arc.nasa.gov, August 9, 2013
http://astrobiology2.arc.nasa.gov/ask-an-astrobiologist/intro/nibiru-and-doomsday-2012-questions-and-answers

"Survive 2012," survive2012.com (Archived)
https://survive2012.com/

Krupp, E.C., "The Great 2012 Scare," skyandtelescope.com, November 2009
https://web.archive.org/web/20130312062008/http://media.skyandtelescope.com/documents/Doomsday2012-lores.pdf

Morrison, David, "The Myth of Nibiru and the End of the World in 2012," www.csicop.org, September 2008
https://web.archive.org/web/20150924035216/http://www.csicop.org/si/show/myth_of_nibiru_and_the_end_of_the_world_in_2012

END TIMES—HEAVEN'S GATE AND THE HALE-BOPP COMET

"Heaven's Gate—How and When It May Be Entered"
https://www.heavensgate.com/

"CNN—Mass Suicide Involved Sedatives, Vodka and Careful Planning," March 27, 1997
http://www.cnn.com/US/9703/27/suicide/index.html

Bearman, Joshuah, "Heaven's Gate: The Sequel," LA Weekly, March 21, 2007
https://www.laweekly.com/heavens-gate-the-sequel/

Reimann, Matt, "Suicide, Nikes, and Comet Space Ships: The Story of the Heaven's Gate Cult." Medium, May 14, 2018
https://timeline.com/the-heavens-gate-mass-suicide-7f440ab4b333

Zeller, Benjamin E., "Anatomy of a Mass Suicide: The Dark, Twisted Story behind a UFO Death Cult." Salon, November 15, 2014
https://www.salon.com/2014/11/15/anatomy_of_a_mass_suicide_the_dark_twisted_story_behind_a_ufo_death_cult/

JIM JONES AND THE JONESTOWN MASSACRE

Conroy, J Oliver, "An Apocalyptic Cult, 900 Dead: Remembering the Jonestown Massacre, 40 Years On," the Guardian, November 21, 2018
https://www.theguardian.com/world/2018/nov/17/an-apocalyptic-cult-900-dead-remembering-the-jonestown-massacre-40-years-on

Collins, John, "The 'Full Gospel' Origins of Peoples Temple—Alternative Considerations of Jonestown & Peoples Temple," September 17, 2019
https://jonestown.sdsu.edu/?page_id=92702

Fondakowski, Leigh, Stories from Jonestown, 2013, ISBN-13 978-0816678082

"Paranoia and Delusions," time.com, December 11, 1978
https://web.archive.org/web/20090114212817/http://www.time.com/time/magazine/article/0,9171,919897-1,00.html

KENYA STARVATION CULT

Mayaka, Emeka, "Mackenzie: Cult Leader Who Led Legions to Death," People Daily, April 25, 2023
https://www.pd.co.ke/news/mackenzie-cult-leader-who-led-legions-to-death-178492/

Mtalaki, Francis, "Shakahola Death Toll Climbs to 372 after 12 More Bodies Exhumed," Citizen Digital, October 31, 2023
https://www.citizen.digital/news/shakahola-death-toll-climbs-to-372-after-12-more-bodies-exhumed-n323399

"Kenya Cult: Children Targeted to Die First, Pastor Says," BBC News, May 14, 2023
https://www.bbc.com/news/world-africa-65588273

Mwai, Peter, Deka Barrow and Rose Njoroge, "Pastor Paul Mackenzie: What Did the Starvation Cult Leader Preach?," BBC News, May 10, 2023
https://www.bbc.com/news/world-africa-65412822

Thiong'o, Josphat, "Kindiki: Makenzi Hired Armed Gangs to Kill Followers Who Took Too Long to Die," The Standard, May 26, 2023
https://www.standardmedia.co.ke/article/2001473797/kindiki-makenzi-hired-armed-gangs-to-kill-followers-who-took-too-long-to-die

NOTES AND SOURCES

LIGHTHOUSE CULT

Mangan, Lucy, "A Very British Cult Review—an Unrelenting Investigation into the Worst of Humanity," the Guardian, April 5, 2023
https://www.theguardian.com/tv-and-radio/2023/apr/05/a-very-british-cult-review-lighthouse-bbc-documentary

"Mentoring and Coaching Company Shut down for Financial Irregularities," www.gov.uk, April 6, 2023
https://www.gov.uk/government/news/mentoring-and-coaching-company-shut-down-for-financial-irregularities

Lighthouse International
https://lighthouseinternationalgroup.com/

BRANCH DAVIDIANS

Gallagher, Eugene V., "Davidians and Branch Davidians (1929-1981)—WRSP," October 8, 2016
https://wrldrels.org/2016/10/08/davidians-and-branch-davidians/

"Report to the Deputy Attorney General on the Events at Waco, Texas: Child Abuse | DOJ | Department of Justice," April 19, 1993
https://web.archive.org/web/20170511015541/https://www.justice.gov/publications/waco/report-deputy-attorney-general-events-waco-texas-child-abuse

Burton, Tara Isabella, "The Waco Tragedy: 30 Years Later, We Still Don't How to Talk about It." Vox, March 23, 2023
https://www.vox.com/2018/4/19/17246732/waco-tragedy-explained-david-koresh-mount-carmel-branch-davidian-cult-30-year-anniversary

FALUN GONG

Zadrozny, Brandy and Ben Collins, "Trump, QAnon and an Impending Judgment Day: Behind the Facebook-Fueled Rise of The Epoch Times," NBC News, August 20, 2019
https://www.nbcnews.com/tech/tech-news/trump-QAnon-impending-judgment-day-behind-facebook-fueled-rise-epoch-n1044121

Cook, Sarah, "Falun Gong: Religious Freedom in China," Freedom House
https://freedomhouse.org/report/2017/battle-china-spirit-falun-gong-religious-freedom

"PEOPLE'S REPUBLIC OF CHINA Crackdown on Falun Gong and Other so-Called 'Heretical Organizations,'" Amnesty International, March 23, 2000
https://www.amnesty.org/en/wp-content/uploads/2021/12/ASA170112000ENGLISH.pdf

Elks, Sonia, "China Is Harvesting Organs from Falun Gong Members, Finds Expert Panel." reuters.com, June 17, 2019
https://www.reuters.com/article/us-britain-china-rights/china-is-harvesting-organs-from-falun-gong-members-finds-expert-panel-idUSKCN1TI236

Gartenberg, Chaim, "Epoch Times Banned from Advertising on Facebook after Breaking Ad Rules," The Verge, August 23, 2019
https://www.theverge.com/2019/8/23/20830229/epoch-times-facebook-ban-buying-ads-violating-policies-trump-propaganda

Perrone, Alessio, "A Key Source for Covid-Skeptic Movements, the Epoch Times Yearns for a Global Audience," Coda Story, April 22, 2022
https://www.codastory.com/disinformation/epoch-times/

Sveen, Benjamin, Lisa McGregor, Hagar Cohen, Eric Campbell, and Matt Henry, "Insiders Reveal the Opaque World of Falun Gong," ABC News, July 31, 2020
https://www.abc.net.au/news/2020-07-21/inside-falun-gong-master-li-hongzhi-the-mountain-dragon-springs/12442518

Carlson, Peter, "For Whom the Gong Tolls," Washington Post, February 27, 2000
https://www.washingtonpost.com/archive/lifestyle/2000/02/27/for-whom-the-gong-tolls/bab9382d-0b90-44da-b4ae-cef517460652/

INTELLIGENT DESIGN AND THE FLYING SPAGHETTI MONSTER

Henderson, Bobby, "Spaghetti Monster—Church of the Flying Spaghetti Monster"
https://spaghettimonster.com/

Durando, Jessica, "Pastafarian Can Wear Strainer on Head in License Photo." USA TODAY, November 16, 2015
https://www.usatoday.com/story/news/nation-now/2015/11/16/church-flying-spaghetti-monster-massachusetts-religion/75862946/

Henderson, Bobby, "Open Letter To Kansas School Board at Church of the Flying Spaghetti Monster,"
https://web.archive.org/web/20070407182624/http://www.venganza.org/about/open-letter/

POLITICAL DELUSIONS

QANON, PROUD BOYS, JANUARY 6 AND THE TRUMP DELUSION

"QAnon," adl.org, December 14, 2022
https://www.adl.org/resources/backgrounder/QAnon

"5 Facts about the QAnon Conspiracy Theories," Pew Research Center, December 15, 2020
https://www.pewresearch.org/short-reads/2020/11/16/5-facts-about-the-QAnon-conspiracy-theories/

Klepper, David, and Ali Swenson, "Trump Openly Embraces, Amplifies QAnon Conspiracy Theories," AP News, September 20, 2022
https://apnews.com/article/technology-donald-trump-conspiracy-theories-government-and-politics-db50c6f709b1706886a876ae6ac298e2

Zadrozny, Brandy and Ben Collins "Who Is behind the QAnon Conspiracy? We've Traced It to Three People," NBC News, August 14, 2018
https://www.nbcnews.com/tech/tech-news/how-three-conspiracy-theorists-took-q-sparked-QAnon-n900531

NOTES AND SOURCES

Collins, Ben and Joe Murphy "Russian Troll Accounts Purged by Twitter Pushed QAnon, Other Conspiracies," NBC News, February 2, 2019
https://www.nbcnews.com/tech/social-media/russian-troll-accounts-purged-twitter-pushed-QAnon-other-conspiracy-theories-n966091

Shanahan, James, "Support for QAnon Is Hard to Measure—and Polls May Overestimate It," The Conversation,
https://theconversation.com/support-for-QAnon-is-hard-to-measure-and-polls-may-overestimate-it-156020

Argentino, Marc-André, "QAnon and the Storm of the U.S. Capitol: The Offline Effect of Online Conspiracy Theories," The Conversation
https://theconversation.com/QAnon-and-the-storm-of-the-u-s-capitol-the-offline-effect-of-online-conspiracy-theories-152815

Rothschild, Mike, "Here Is Every QAnon Prediction That's Failed to Come True," The Daily Dot, January 27, 2021
https://www.dailydot.com/debug/QAnon-failed-predictions/

Marina Pitofsky, "QAnon Supporters Gather over Theory That JFK Jr. Will Emerge, Announce Trump to Be Reinstated." USA TODAY, November 2, 2021
https://www.usatoday.com/story/news/nation/2021/11/02/texas-QAnon-believers-back-theory-trump-reinstated/6255234001/

LaCapria, Kim, "Is Comet Ping Pong Pizzeria Home to a Child Abuse Ring Led by Hillary Clinton?" Snopes, November 21, 2016
https://www.snopes.com/fact-check/pizzagate-conspiracy/

"Proud Boys," Southern Poverty Law Center
https://www.splcenter.org/fighting-hate/extremist-files/group/proud-boys

Sardarizadeh, Shayan, and Alistair Coleman "Biden Inauguration: What Are Far-Right Trump Supporters Saying?," BBC News, January 19, 2021
https://www.bbc.com/news/blogs-trending-55679813

Johnson, Ted, "Donald Trump Becomes First President To Be Impeached Twice; House Charges Him With Inciting Capitol Hill Siege" Deadline, January 13, 2021
https://deadline.com/2021/01/donald-trump-impeachment-house-of-representatives-1234672714/

Hassan, Steve, "The Cult of Trump—A Leading Cult Expert Explains How Trump Uses Mind Control," Free Press, October 15, 2019

DOMINION VOTING MACHINES

Rieder, Rem, "Trump Tweets Conspiracy Theory About Deleted Votes,"FactCheck.Org, November 13, 2020
https://www.factcheck.org/2020/11/trump-tweets-conspiracy-theory-about-deleted-votes/https://www.factcheck.org/2020/11/trump-tweets-conspiracy-theory-about-deleted-votes/

Giles, Christopher and Jake Horton, "US Election 2020: Is Trump Right about Dominion Machines?" BBC News, November 17, 2020
https://www.bbc.com/news/election-us-2020-54959962

NOTES AND SOURCES

Thorbecke, Catherine, Mike Hayes, Maureen Chowdhury, Marshall Cohen, Oliver Darcy, Jon Passantino, Elise Hammond and Tori B. Powell, "Settlement Reached in Dominion Defamation Lawsuit against Fox News," cnn.com, April 19, 2023
https://www.cnn.com/business/live-news/fox-news-dominion-trial-04-18-23

Zeidman, Bob "How I Won $5 Million From the MyPillow Guy and Saved Democracy," POLITICO, May 26, 2023
https://www.politico.com/news/magazine/2023/05/26/my-pillow-mike-lindell-investigation-00097903

SUICIDE BOMBING

Hoffman, Bruce, "The Logic of Suicide Terrorism," The Atlantic, June 6, 2018
https://www.theatlantic.com/magazine/archive/2003/06/the-logic-of-suicide-terrorism/302739/

"Suicide Bombings: What Does the Law Actually Say?'" aoav.org, June 5, 2015
https://web.archive.org/web/20150611185742/https://aoav.org.uk/2015/suicide-bombings-what-does-the-law-say/

"A Short History of Suicide Bombing," aoav.org, October 20, 2013
https://web.archive.org/web/20140125110530/https://aoav.org.uk/2013/a-short-history-of-suicide-bombings/

Harmon, Vanessa, Edin Mujkic, Catherine Kaukinen, and Henriikka Weir, "Causes and Explanations of Suicide Terrorism: A Systematic Review," Homeland Security Affairs 14, Article 9, December 2018,
https://www.hsaj.org/articles/14749

THE KIM DYNASTY OF NORTH KOREA

"Next of Kim," The Economist, December 22, 2011
https://www.economist.com/asia/2010/09/23/next-of-kim

"North Korea's Secretive 'First Family,'" BBC News, December 13, 2013
https://www.bbc.com/news/world-asia-pacific-11297747

Demick, Barbara, "Nothing to Envy," Spiegel & Grau, 2009

"Kim Family | North Korea Leadership Watch"
https://www.nkleadershipwatch.org/kim-family/

"Kim Jong-Nam Says N.Korean Regime Won't Last Long," The Chosun Ilbo (English Edition): Daily News from Korea—North Korea, January 17, 2012
https://english.chosun.com/site/data/html_dir/2012/01/17/2012011701790.html

"Korea, North—The World Factbook," cia.gov
https://www.cia.gov/the-world-factbook/countries/korea-north/

Hotham, Oliver, "The Weird, Weird World of North Korean Elections." North Korea News, March 3, 2014
https://www.nknews.org/2014/03/the-weird-weird-world-of-north-korean-elections/

Fisher, Max, "The Single Most Important Fact for Understanding North Korea," Vox, January 6, 2016
https://www.vox.com/2016/1/6/10724334/north-korea-history

Harden, Blaine, "Outside World Turns Blind Eye to N. Korea's Hard-Labor Camps," July 20, 2009
https://www.washingtonpost.com/wp-dyn/content/article/2009/07/19/AR2009071902178.html

"The State of North Korean Farming: New Information from the UN Crop Assessment Report," 38 North, February 24, 2017
https://www.38north.org/2013/12/rireson121813/

"Life Inside North Korea," U.S. Department of State, June 5, 2003
https://2001-2009.state.gov/p/eap/rls/rm/2003/21269.htm

Schielke, Thomas, "How Satellite Images of the Earth at Night Help Us Understand Our World and Make Better Cities," ArchDaily, July 29, 2018
https://www.archdaily.com/892730/how-satellite-images-of-the-earth-at-night-help-us-understand-our-world-and-make-better-cities

ALBANIA—THE HERMIT KINGDOM

"Albania—The World Factbook," cia.gov
https://www.cia.gov/the-world-factbook/countries/albania/

Fischer, Bernd, "Albania and Enver Hoxha's Legacy," openDemocracy.net, June 10, 2010
https://www.opendemocracy.net/en/albania-and-enver-hoxhas-legacy/

"Albania," Freedom House.org
https://freedomhouse.org/country/albania/freedom-world/2023

White, Jeffrey, "Letter from Albania: Enver Hoxha's Legacy, and the Question of Tourism" Gadling.com, June 26, 2008
https://gadling.com/2008/06/26/letter-from-albania-enver-hoxhas-legacy-and-the-question-of-t/

Wilkinson, Chris, "Suspicious Minds—Enver Hoxha & Albania: A Cult of Capriciousness," June 28, 2020
https://www.linkedin.com/pulse/suspicious-minds-enver-hoxha-albania-cult-chris-wilkinson/

Sakalis, Alex, "Enver Hoxha: The Lunatic Who Took Over the Asylum," March 23, 2016
https://isnblog.ethz.ch/uncategorized/enver-hoxha-the-lunatic-who-took-over-the-asylum

THE COMMUNIST DELUSION

Ball, Terence, and Richard Dagger, "Communism | Definition, History, Varieties, & Facts," Encyclopedia Britannica, July 26, 1999
https://www.britannica.com/topic/communism

Karlsson, Klas-Göran, and Michael Schoenhals, "Crimes against Humanity under Communist Regimes," The Living History Forum, 2008, ISBN: 978-91-977487-2-8
https://www.levandehistoria.se/sites/default/files/material_file/research-review-crimes-against-humanity.pdf

Greaves, Bettina Bien, "Why Communism Failed," Foundation for Economic Education, March 1, 1991
https://fee.org/articles/why-communism-failed/

NOTES AND SOURCES

Piereson, James, "Socialism as a Hate Crime" The New Criterion," August 21, 2018
  https://newcriterion.com/blogs/dispatch/socialism-as-a-hate-crime-9746

Satter, David, "100 Years of Communism—and 100 Million Dead," WSJ, November 6, 2017
  https://www.wsj.com/articles/100-years-of-communismand-100-million-dead-1510011810

Puddington, Arch, "In Modern Dictatorships, Communism's Legacy Lingers On," Freedom House, March 23, 2017
  https://freedomhouse.org/article/modern-dictatorships-communisms-legacy-lingers

Made in the USA
Columbia, SC
23 December 2024

50541497R00154